DMV Seminar
Band 6

Birkhäuser Verlag
Basel · Boston · Stuttgart

A. Delgado
D. Goldschmidt
B. Stellmacher

Groups and Graphs: New Results and Methods

1985

Birkhäuser Verlag
Basel · Boston · Stuttgart

Authors

A. Delgado
Dept. of Mathematics
Kansas State University
Manhattan, KS 6650
USA

D. Goldschmidt
Dept. of Mathematics
University of California at
Berkeley
Berkeley, Cal. 94720
USA

B. Stellmacher
Fakultät für Mathematik
Universität Bielefeld
Universitätsstrasse
D–4800 Bielefeld

The seminar was made possible through the support of the *Stiftung Volkswagenwerk.*

Library of Congress Cataloging in Publication Data

Delgado, A. (Alberto)
Groups and graphs.
(DMV Seminar ; Bd. 6)
1. Groups, Theory of. 2. Graph theory.
I. Goldschmidt, David M. II. Stellmacher, B.
(Bernd) III. Title. IV. Series.
QA171.D4 1985 511'.5 85–15644
ISBN 3-7643-1736-1

CIP-Kurztitelaufnahme der Deutschen Bibliothek

Delgado, Alberto:
Groups and graphs: new results and methods / A. Delgado ;
D. Goldschmidt ; B. Stellmacher. – Basel ; Boston ;
Stuttgart : Birkhäuser, 1985.
(DMV-Seminar ; Bd. 6)
ISBN 3-7643-1736-1
NE: Goldschmidt, David ; Stellmacher, Bernd ; Deutsche
Mathematiker-Vereinigung: DMV-Seminar

Printed in Germany
ISBN 3-7643-1736-1

Preface.

This set of notes began as a report on the seminar "Gruppen und Graphen" organized by the Deutsche Mathemaitker-Vereinigung and held in Düsseldorf from 20.6. - 25.6.1982. We have greatly expanded the material presented there. The text is divided into two main sections. The first part is an exposition of the Bass-Serre theory of groups and graphs written by D.M. Goldschmidt. The second part, a joint work of A.L. Delgado and B. Stellmacher, presents the classification of "weak (B,N)-pairs of rank 2". A special case of their result was delivered at the seminar.

The prerequisites for reading this book are basic group theory and a certain degree of mathematical sophistication. It was not our intention to present an exhaustive account of the research in graphs and groups, but rather to give an overview of the techniques one uses in their study. We wanted, on the one hand, to give the non-specialist an impression of the more classical aspects of this field and, on the other hand, to give those, beginning work in this field, a fairly detailed picture of the latest tools used here.

We hope in this way to accomplish what we set out to do in 1982. We want to thank the DMV and in particular Professor Dr. Gerd Fischer for giving us the opportunity to present this work to a larger audience and especially to younger mathematicians. We also thank the "Stiftung Volkswagenwerk" for their generous support in making the conference possible. Thanks go especially to the participants of the conference for their lively interest.

In the preparation of this manuscript we were helped very much by Thomas Westerhoff. Applause also goes to Frau Marlies Fettköther for her valiant effort at the typewriter.

Bielefeld, April 1985.

Alberto Delgado, Kansas State University
Bernd Fischer, Universität Bielefeld
David Goldschmidt, University of California at Berkeley
Bernd Stellmacher, Universität Bielefeld

Contents

Part I

Graphs and Groups

D.M. Goldschmidt

Part II

Weak (B,N)-pairs of rank 2

A.L. Delgado and B. Stellmacher

Part I

Graphs and Groups

D. M. Goldschmidt

1. Introduction.

In this first part of the book, we give an exposition of the Bass-Serre theory of groups acting on graphs [3]. Our approach is based on a clever idea of Warren Dicks [1]. The main objective is to obtain generators and relations for a group G acting on a connected graph Γ in terms of the isotropy subgroups of G and the fundamental group of Γ. When Γ is a tree, one gets an explicit presentation.

As a by-product, more or less, of the analysis, it follows that an abstract group given by a certain type of presentation (e.g. free product with amalgamation) acts naturally on a tree. From this, one can easily obtain "normal form theorems" and a subgroup classification.

2. Definitions, notations, and preliminary results.

By a graph Γ we shall mean the disjoint union of two sets $E(\Gamma) \dot{\cup} V(\Gamma)$ together with two "incidence" maps from $E(\Gamma)$ to $V(\Gamma)$ which we shall call $+1$ and -1. The elements of $E(\Gamma)$ are called edges and those of $V(\Gamma)$ vertices. For $e \in E(\Gamma)$, we write e^{+1}, e^{-1} for the two vertices of e. The edge is said to be oriented and is depicted

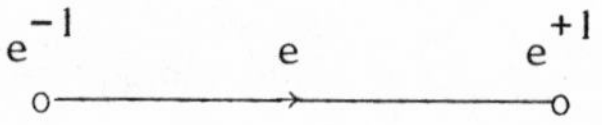

An edge e with $e^{-1} = e^{+1}$ is called a loop. Let Γ and Γ' be graphs, then a map $\psi : \Gamma \longrightarrow \Gamma'$ is a pair of functions $\psi_E : E(\Gamma) \longrightarrow E(\Gamma')$ and $\psi_V : V(\Gamma) \longrightarrow V(\Gamma')$ which preserve the incidence maps. The space of paths $P(\Gamma)$ is the free abelian group on the set $E(\Gamma)$. A path γ is an element $\gamma = \sum_{i=1}^{n} \varepsilon_i e_i$ of $P(\Gamma)$ such that $\varepsilon_i = \pm 1$ for all i, $e_i^{\varepsilon_i} = e_{i+1}^{-\varepsilon_{i+1}}$ $(1 \leq i < n)$ and $e_i \neq e_{i+1}$ $(1 \leq i < n)$. The path γ then has length n, denoted $|\gamma| = n$. Define $\gamma^{+1} = e_n^{\varepsilon_n}$ and $\gamma^{-1} = e_1^{-\varepsilon_1}$. We call γ a circuit if $\gamma^{+1} = \gamma^{-1}$ and we say that two vertices $v, w \in V(\Gamma)$ are connected if $v = w$ or if there is a path γ with $\gamma^{-1} = v$ and $\gamma^{+1} = w$. The resulting equivalence relation partitions Γ into connected components in an obvious way. A tree is a connected graph with no circuits. For a connected graph Γ let $d : V(\Gamma) \times V(\Gamma) \to \mathbb{Z}$ be the metric on Γ given by:

$$d(v,w) = \begin{cases} 0 & \text{if } v = w, \\ \min\{|\gamma| \mid \gamma^{+1} = v,\ \gamma^{-1} = w\} & \text{if } v \neq w. \end{cases}$$

For $i \in \mathbb{N}$, $\alpha \in V(\Gamma)$ let $\Delta^{(i)}(\alpha) = \{\beta \in V(\Gamma) \mid d(\alpha,\beta) \leq i\}$. We write simply $\Delta(\alpha)$ for $\Delta^{(1)}(\alpha)$.

Let G be a group. A <u>right</u> (<u>left</u>) G-<u>set</u> is a set X with a map $X \times G \longrightarrow X$ $(G \times X \longrightarrow X)$ written $x \longrightarrow x^g$ $(x \longrightarrow {}^g x)$ such that $x^1 = x$ $({}^1x = x)$ and $(x^{g_1})^{g_2} = x^{g_1g_2}$ $({}^{g_1}({}^{g_2}x) = {}^{g_1g_2}x)$ for all $g_1, g_2 \in G$. The term G-set will mean right G-set. The <u>quotient set</u> X/G $(G\backslash X)$ is the set of equivalence classes under the relation $x \sim x^g$ $(x \sim {}^g x)$. When there is no confusion we may write $x \longrightarrow \bar{x}$ for the natural map of X onto X/G $(G\backslash X)$. The elements of X/G $(G\backslash X)$ are called G-<u>orbits</u>. If H is a subgroup of G, then G is both a left and right H-set with ${}^h g = hg$ and $g^h = gh$ respectively, and $H\backslash G$ (G/H) is a right (left) G-set. It follows that $|(H\backslash G)/G| = 1 = |G\backslash(G/H)|$. An <u>isomorphism</u> of right (left) G-sets X,Y is a bijective map $\varphi : X \longrightarrow Y$ with $\varphi(x^g) = \varphi(x)^g$ $(\varphi({}^g x) = {}^g\varphi(x))$ for all $x \in X$, all $g \in G$. A <u>transversal</u> for a right (left) G-set is a choice function $t : X/G \longrightarrow X$ $(G\backslash X \longrightarrow X)$ with $t(\bar{y}) \in \bar{y}$. The <u>fundamental theorem of</u> G-<u>sets</u> says that given a transversal $t : X/G \longrightarrow X$ there is an isomorphism of G-sets $X \simeq \bigcup_{y \in X/G} G_{t(y)}\backslash G$, where $G_x = \{g \in G \mid x^g = x\}$. For further reference, we record:

(2.1) <u>Let</u> X <u>be a</u> G-<u>set and</u> $t : X/G \longrightarrow X$ <u>a transversal</u>. <u>Then there is an isomorphism</u> $\bigcup_{y \in X/G} G_{t(y)}\backslash G \simeq X$ <u>given by</u> $G_{t(y)}g \longrightarrow t(y)^g$ <u>for all</u> $g \in G$, $y \in X/G$. □

We will, in the following, be dealing with several categories and functors between them. The reader unfamiliar with this language should have nothing to fear. It allows us to treat widely differing concepts in a unified way. After a little practice it becomes quite natural. A good reference is [2].

For any category $\mathcal{C}$ we will denote the objects of $\mathcal{C}$ by $\mathcal{O}(\mathcal{C})$ and the morphisms between two objects X,Y will be denoted by Hom(X,Y). The following two categories will be of particular importance to us.

<u>Example</u> (2.2): Let G be a group. We define a category $\hat{G}$ by taking for $\mathcal{O}(\hat{G})$ the set of subgroups of G. For two objects (subgroups) X,Y in $\hat{G}$ set $\mathrm{Hom}(X,Y) = \{\mathrm{ad}(g) \mid g \in G, \mathrm{ad}(g)X = gXg^{-1} \leq Y\}$.

<u>Example</u> (2.3): Let Γ be a graph. We define a category $\hat{\Gamma}$ by taking for $\mathcal{O}(\hat{\Gamma})$ the vertices and edges of Γ. The non-identity morphisms are simply the incidence maps. More precisely:

$$\mathrm{Hom}(P,P) = \{1_P\}, \quad P \in \Gamma,$$

$$\mathrm{Hom}(P,Q) = \begin{cases} \{(P,Q,+1)\}, & \text{if } P \in E(\Gamma),\ Q \in V(\Gamma),\ Q = P^{+1} \\ \{(P,Q,-1)\}, & \text{if } P \in E(\Gamma),\ Q \in V(\Gamma),\ Q = P^{-1} \\ \emptyset, & \text{otherwise.} \end{cases}$$

3. G-graphs.

Let Γ be a graph and G be a group. We say that Γ is a G-graph if $E(\Gamma)$ and $V(\Gamma)$ are G-sets and the action of G preserves incidence: $(e^{\pm 1})^g = (e^g)^{\pm 1}$ for all $e \in E(\Gamma)$, all $g \in G$. We then define the quotient graph Γ/G with $E(\Gamma/G) = E(\Gamma)/G$, $V(\Gamma/G) = V(\Gamma)/G$, and $\bar{e}^{\pm 1} = \overline{(e^{\pm 1})}$ for all $e \in E(\Gamma)$.

Suppose $t : \Gamma/G \longrightarrow \Gamma$ is a transversal. Then for each $y \in E(\Gamma/G)$ there is a pair of group elements $t_{\pm 1}(y)$ such that

$$(3.1) \qquad t(y^{\pm 1})^{t_{\pm 1}(y)} = t(y)^{\pm 1}.$$

A transversal $t : \Gamma/G \longrightarrow \Gamma$ together with such a pair of choice functions $t_{\pm 1} : E(\Gamma/G) \longrightarrow G$ will be called an augmented transversal for Γ/G.

We shall illustrate the meaning of equation (3.1) with the following example: Let Γ be the graph

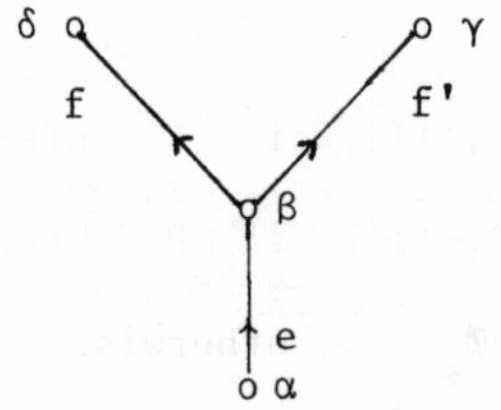

The group $G = \mathbb{Z}_2$ operates on Γ with factor graph Γ/G

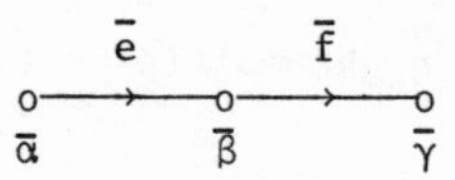

Choose a transversal t with $t(\bar{\alpha}) = \alpha$, $t(\bar{\beta}) = \beta$, $t(\bar{\gamma}) = \gamma$ and $t(\bar{e}) = e$, $t(\bar{f}) = f$. The transversal is "good" with respect to the edge $\bar{e}$, for $t(\bar{e}^{\pm 1}) = t(\bar{e})^{\pm 1}$, but it is "bad" with respect to $\bar{f}$, because $t(\bar{f}^{+1}) \neq t(\bar{f})^{+1}$. However, as f and f' are in the same G-orbit there exists an element $t_{+1}(f)$ of the operating group G such that $\gamma^{t_{+1}(f)} = t(\bar{f}^{+1})^{t_{+1}(f)} = t(\bar{f})^{+1} = \delta$, and (3.1) is satisfied.

One of our aims will be to find a way to choose a "good" transversal that allows us to take $t_{\pm 1}(y) = 1$ for most $y \in E(\Gamma/G)$.

Now an augmented transversal $t : \Gamma/G \longrightarrow \Gamma$ allows us to recover the graph Γ from the group G. Namely, we define a graph $\tilde{\Gamma}$ with $E(\tilde{\Gamma}) = \bigcup_{y \in E(\Gamma/G)} G_{t(y)} \backslash G$, $V(\tilde{\Gamma}) = \bigcup_{y \in V(\Gamma/G)} G_{t(y)} \backslash G$ and

$$(G_{t(y)} g)^{\pm 1} = G_{t(y^{\pm 1})} t_{\pm 1}(y) g$$

for all $y \in E(\Gamma/G)$. Using (3.1) one verifies that the incidence maps for $\tilde{\Gamma}$ are well defined. Then G acts by right-multiplication, and it is easy to check that

(3.2) <u>The isomorphism given by</u> (2.1) <u>is an isomorphism of</u> G-<u>graphs</u> $\Gamma \simeq \tilde{\Gamma}$. □

This fact allows us to describe a G-graph "internally" ; that is, the structure of a G-graph is completely determined by a set of subgroups of G (the vertex- and edge-stabilizers) and the operation of G on the cosets of these subgroups via right-multiplication.

A useful property of the natural map $\Gamma \longrightarrow \Gamma/G$ is that it is <u>locally surjective</u>. Namely, if $\bar{v}$ and $\bar{e}$ are incident in Γ/G, then there must be an edge e' of Γ which is incident to v and is G-conjugate to e. Thus, all edges of Γ/G incident to a vertex $\bar{v}$ can be lifted to edges of Γ which are incident to v.

(3.3) <u>The natural map</u> $\Gamma \rightarrow \Gamma/G$ <u>is locally surjective</u>. <u>In particular</u>, <u>the image of a connected component of</u> Γ <u>is a connected component of</u> Γ/G.

<u>Proof</u>: Let Γ_o be a connected component of Γ. Then $\bar{\Gamma}_o$ is clearly connected. If it were not a component of $\bar{\Gamma}$, we could find a vertex $\bar{v}$ of $\bar{\Gamma}_o$ incident to an edge $\bar{e}$ of $\bar{\Gamma} - \bar{\Gamma}_o$. But this is impossible since $\bar{e}$ can be lifted to an edge of Γ_o by local surjectivity. □

If, in addition to local surjectivity, we have $G_v = 1$ for all $v \in V(\Gamma)$, i.e. <u>free action</u>, then the natural map is a <u>local isomorphism</u> and we sometimes say that Γ <u>covers</u> Γ/G. In this situation we shall make use of the well-known fact that a tree has no proper connected coverings:

(3.4) <u>Suppose that the connected</u> G-<u>graph</u> Γ <u>covers the tree</u> Γ/G. <u>Then</u> $G = 1$.

<u>Proof</u>: Suppose that a connected G-graph Γ covers Γ/G and $g \in G - \{1\}$. Choose $v \in V(\Gamma)$ and let $\gamma = \Sigma\, \varepsilon_i e_i$ be a path connecting v and v^g. Since G is acting freely, $\bar{e}_i \neq \bar{e}_{i+1}$ for any i, and thus $\bar{\gamma} = \Sigma\, \varepsilon_i \bar{e}_i$ is a circuit in Γ/G. □

The above result is useful not so much when applied to the action of G but when applied to the action of a normal subgroup of G. That is, given a G-graph Γ and a normal subgroup N of G, Γ/N becomes a G-graph in the obvious way: $\bar{x}^g = \overline{x^g}$. This is easily seen to be well-defined, and it makes the natural map $\Gamma \to \Gamma/N$ into a map of G-graphs. Viewing Γ/G as a G-graph with trivial action, we get a commutative diagram of G-graphs and natural maps:

(3.5)

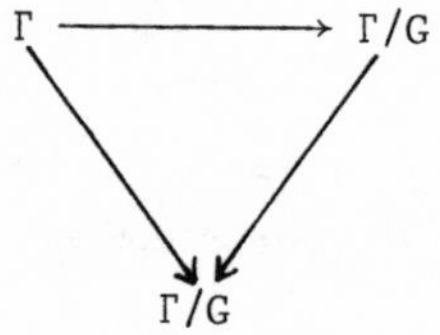

It is useful to consider the following weakening of the notion of connectivity. We shall call a G-graph Γ G-<u>connected</u> if Γ/G is connected. In this situation G is clearly transitive on connected components of Γ, in fact by (3.3) there exists a transversal $t : \Gamma/G \longrightarrow \Gamma$ whose image is contained in a connected component of Γ.

(3.6) <u>Let</u> Γ <u>be a</u> G-<u>connected</u> G-<u>graph</u>. <u>Let</u> Γ_o <u>be a connected component of</u> Γ <u>and let</u> G_o <u>be the (set-wise) stabilizer of</u> Γ_o <u>in</u> G. <u>Let</u> $t : \Gamma/G \longrightarrow \Gamma$ <u>be an augmented transversal with</u> $t(y) \in \Gamma_o$ <u>for all</u> $y \in \Gamma/G$. <u>Then</u> $G_o = \langle G_{t(\bar{v})}, t_{\pm 1}(\bar{e}) \mid \bar{v} \in V(\Gamma/G), \bar{e} \in E(\Gamma/G)\rangle$.

<u>Proof</u>: Put $\tilde{G}_o = \langle G_{t(\bar{v})}, t_{\pm 1}(\bar{e}) \mid \bar{v} \in V(\Gamma/G), \bar{e} \in E(\Gamma/G)\rangle$. Since G permutes connected components of Γ, we get $g \in G_o$ whenever $\Gamma_o \cap \Gamma_o^g \neq \phi$. In particular, $\tilde{G}_o \subseteq G_o$ since $t(y) \in \Gamma_o$ for all $y \in \Gamma/G$.

For $\alpha,\beta \in V(\Gamma_o)$, let $d(\alpha,\beta)$ be the length of the shortest path connecting α to β, and for $g \in G_o$, define

$$d(g) = \min_{\bar{v}_1,\bar{v}_2 \in V(\Gamma/G)} d(t(\bar{v}_1),t(\bar{v}_2)^g).$$

Note that if $d(g) = 0$ then $\bar{v}_1 = \bar{v}_2$ and $g \in G_{t(\bar{v}_1)} \subseteq \tilde{G}_o$. We argue that $g \in \tilde{G}_o$ for any $g \in G_o$ by induction on $d(g)$, having now disposed of the case $d(g) = 0$. Thus, let $\gamma = \sum_{i=1}^{d} \varepsilon_i e_i$ be a path in Γ_o with $d = d(g)$, $\gamma^{-1} = t(\bar{v}_1)$, $\gamma^{+1} = t(\bar{v}_2)^g$ for some $\bar{v}_1,\bar{v}_2 \in \Gamma/G$. We may assume that $v_1 = t(\bar{v}_1)$. Let $v = e_1^{\varepsilon_1}$, then $\sum_{i=2}^{d} \varepsilon_i e_i$ is a path from v to $t(\bar{v}_2)^g$. Choose $x \in G_o$ with $v^x = t(\bar{v})$, then $d(gx) \leq d-1$ so $gx \in \tilde{G}_o$ by induction, and it remains to show that $x \in \tilde{G}_o$. Choose $x_1 \in G_o$ with $e_1^{x_1} = t(\bar{e}_1)$. Now, $\bar{v}_1 = \bar{e}_1^{-\varepsilon_1}$, and $v_1 = t(\bar{v}_1) = t(\bar{e}_1^{-\varepsilon_1})$. Hence

$$v_1^{t_{-\varepsilon_1}(\bar{e}_1)} = t(\bar{e}_1)^{-\varepsilon_1} = (e_1^{-\varepsilon_1})^{x_1} = v_1^{x_1}.$$

We conclude that $t_{-\varepsilon_1}(\bar{e}_1)x_1^{-1} \in G_{v_1} \subseteq \tilde{G}_o$ and hence, that $x_1 \in \tilde{G}_o$.

However, we also have

$$v^{x_1} = t(\bar{e}_1)^{\varepsilon_1} = t(\bar{e}_1^{\varepsilon_1})^{t_{\varepsilon_1}(\bar{e}_1)} = t(\bar{v})^{t_{\varepsilon_1}(\bar{e}_1)}.$$

So we may take $x = x_1 t_{\varepsilon_1}(\bar{e}_1)^{-1} \in \tilde{G}_o$ with $v^x = t(\bar{v})$. □

The remainder of the first part of this book is essentially an attempt to understand the relations satisfied by the generators given in (3.6). But first of all we shall present some examples to illustrate these notions.

Example (3.7): Let Σ_5 be the symmetric group of degree 5. We define an Σ_5-graph Γ as follows: $V(\Gamma)$ is the set of transpositions of Σ_5; that is, elements of the form (ij), $1 \leq i \neq j \leq 5$; $E(\Gamma)$ is given by $\{(v,w) \in V(\Gamma) \times V(\Gamma) \mid vw = wv\}$, and the incidence maps are defined by: $(v,w)^{-1} = v$, $(v,w)^{+1} = w$. We get the so-called Petersen-graph with 10 vertices and 30 edges.

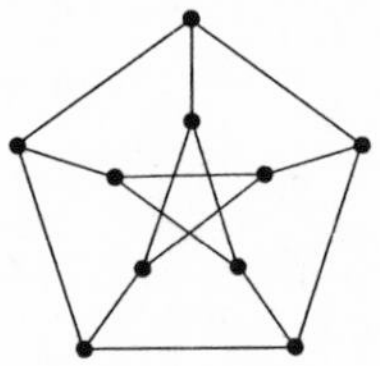

Fig. 1: The Petersen graph

Figure 1 shows the undirected Petersen graph. In order to get the graph described above, replace each edge by two edges with opposite orientation.

Obviously Σ_5 operates via conjugation on Γ and, in fact, transitively on the edges and the vertices of Γ so that the factor graph reduces to a loop:

$\Gamma/\Sigma_5 : \bar{y}$ $\bar{e}$.

Set $G = \Sigma_5$, $\bar{\Gamma} = \Gamma/G$. We recover the original graph from the factor graph with, for example, the following transversal $t : \bar{\Gamma} \longrightarrow \Gamma : t(\bar{y}) = (12)$, $t(\bar{e}) = ((35),(24))$. This is really a "bad" transversal, thus, we have to augment it with $t_{-1}(\bar{e}) = (13)(25)$, $t_{+1}(\bar{e}) = (124)$. Then $G_{t(\bar{y})} = C_G((12))$ is isomorphic to a dihedral group of order 12 and $G_{t(\bar{e})} = C_G(<(35),(24)>)$ is elementary abelian of order 4. It is now easy to check (3.2).

<u>Example</u> (3.8): Let V be a vector space of dimension 3 over GF(2). We take the one- and two-dimensional subspaces of V as the vertices of a graph Γ ; let U be a one-dimensional and W a two-dimensional subspace of V, then $(U,W) \in E(\Gamma)$ if U is a subspace of W, and we define the incidence maps as follows $(U,W)^{-1} = U$, $(U,W)^{+1} = W$. We get a graph, the incidence graph of V, with 14 vertices and 21 edges:

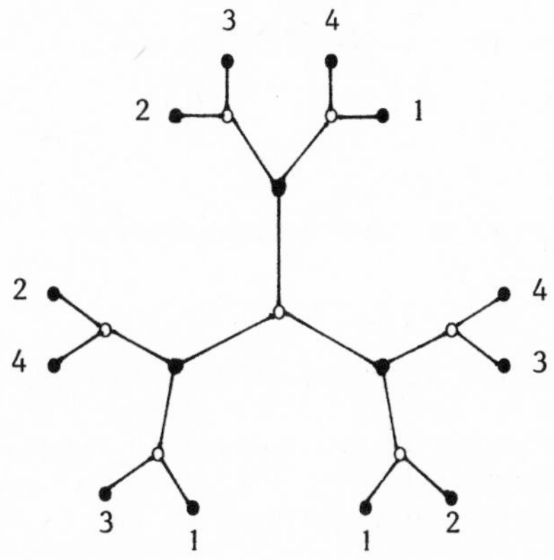

Fig. 2: The incidence graph of V

In the above figure identify the vertices with the same number and note that the hollow nodes are the one-dimensional subspaces, the dark ones the two-dimensional subspaces.

Aut(V) = $GL_3(2)$ operates on Γ, and, in fact, transitively on the edges and on each of the two types of vertices. We get the following factor graph:

$\Gamma/GL_3(2)$: 

As we have seen in the preceding example, we only need to find the stabilizers of the vertices and the edges in order to internally describe the graph Γ. One checks that the stabilizer of each type of vertex is isomorphic to Σ_4 and that an edge stabilizer is isomorphic to a dihedral group of order 8. Now recover the graph Γ from the factor graph and vertex- and edge-stabilizers as above.

Let (,) be the standard symmetric inner product on V and let $\sigma : V(\Gamma) \longrightarrow V(\Gamma)$ be the map given by:

$$\sigma(U) = U^{\perp} .$$

Then it is easy to check that this map induces an action on Γ' (Γ' being derived from Γ by adding edges of the form (W,U) to Γ with W,U as above). This action interchanges the two types of vertices and setting $\Gamma L_3(2) = \langle GL_3(2),\sigma\rangle$ we have that $\Gamma'/\Gamma L_3(2)$ is a loop as in the first example.

4. The universal realization of a graph of groups.

We wish to begin by abstracting the data which is necessary to define generators as in (3.6). For this purpose it is convenient to view the vertices and edges of a graph Y as objects of a category $\hat{Y}$ whose non-identity morphisms are precisely the incidence maps. This was discussed as Example (2.3). By a <u>graph of groups</u> $(\mathcal{Y},Y)$ we mean a functor $\mathcal{Y}$ from $\hat{Y}$ to the category of groups. Thus, for each $y \in Y$ there is a group $\mathcal{Y}(y)$ and for each $e \in E(Y)$ there is a pair of group homomorphisms $\mathcal{Y}_{\pm 1}(e) : \mathcal{Y}(e) \longrightarrow \mathcal{Y}(e^{\pm 1})$. This can be suggestively depicted:

$$\mathcal{Y}(e^{-1}) \xleftarrow{\;\mathcal{Y}_{-1}(e)\;} \mathcal{Y}(e) \xrightarrow{\;\mathcal{Y}_{+1}(e)\;} \mathcal{Y}(e^{+1})$$

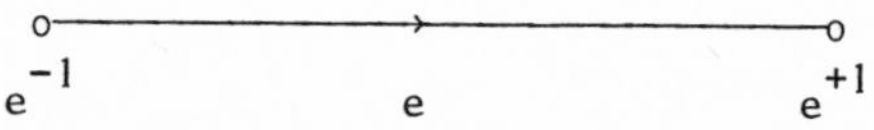

Given a G-graph Γ and an augmented transversal $t : \Gamma/G \longrightarrow \Gamma$, we can obtain a graph of groups in a natural way. Namely, let $Y = \Gamma/G$, $\mathcal{Y}(y) = G_{t(y)}$ and $\mathcal{Y}_{\pm 1}(e) = \text{ad } t_{\pm 1}(e)$ where $\text{ad}(x)$ denotes the inner automorphism $g \longmapsto xgx^{-1}$. (To be more precise, $\mathcal{Y}_{\pm 1}(e)$ is the restriction of $\text{ad } t_{\pm 1}(e)$ to $\mathcal{Y}(e)$). One checks from the defining equation (3.1) that $\mathcal{Y}_{\pm 1}(e)$ are a pair of homomorphisms from $\mathcal{Y}(e)$ to $\mathcal{Y}(e^{\pm 1})$. Later on, we shall see that every graph of groups arises in this way, provided only that the maps $\mathcal{Y}_{\pm 1}(e)$ are monic.

Notice that in the above construction we not only have a graph of groups, but a particular "realization" of it by means of subgroups and inner automorphisms of a group G. In order to formalize this notion for any group G we denote by $\hat{G}$ the category derived from G as in

Example (2.2). There is an obvious functor from this category to the category of groups, which we shall call $\mathcal{A}d$. Thus, if $\mathcal{R}$ is a functor from the category $\hat{Y}$ to $\hat{G}$, we write $\mathcal{R} : \hat{Y} \longrightarrow \hat{G}$, then $(\mathcal{A}d \circ \mathcal{R}, Y)$ is a graph of groups. By a <u>realization of a graph of groups</u> $(\mathcal{Y}, Y)$ <u>in a group</u> G, we mean a pair $(\mathcal{R}, \rho)$ where $\mathcal{R} : \hat{Y} \longrightarrow \hat{G}$ and $\rho : \mathcal{R} \longrightarrow \mathcal{A}d \circ \mathcal{R}$ is a natural transformation of functors. What this means in plain english is that for each $y \in Y$ there is a subgroup $\mathcal{Y}(y)$ of G, and a homomorphism $\rho_y : \mathcal{Y}(y) \longrightarrow \mathcal{R}(y)$. For each $e \in E(Y)$ there is a pair of elements $\mathcal{R}_{\pm 1}(e) \in G$ such that

$$\mathcal{R}_{\pm 1}(e)\ \mathcal{R}(e)\ \mathcal{R}_{\pm 1}(e)^{-1} \subseteq \mathcal{R}(e^{\pm 1}) ,$$

and such that all diagrams of the form

$$\begin{array}{ccc} \mathcal{Y}(e) & \xrightarrow{\ \mathcal{Y}_{\pm 1}(e)\ } & \mathcal{Y}(e^{\pm 1}) \\ \rho_e \downarrow & & \downarrow \rho_{e^{\pm 1}} \\ \mathcal{R}(e) & \xrightarrow[\ \mathrm{ad}\ \mathcal{R}_{\pm 1}(e)\ ]{} & \mathcal{R}(e^{\pm 1}) \end{array} \tag{4.1}$$

commute.

Of particular interest is the case that ρ_y is an isomorphism for all $y \in Y$. We shall say that $(\mathcal{R}, \rho)$ is a <u>faithful realization</u> in this case. We shall often abuse notation by saying that "$\mathcal{R}$ is a realization of $(\mathcal{Y}, Y)$ in G".

Given a realization $\mathcal{R}$ of a graph of groups $(\mathcal{Y}, Y)$ in G, we next define the graph of the realization, $\Gamma(\mathcal{R})$, with

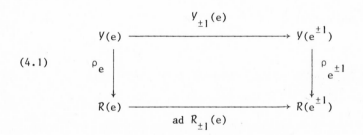

$$E(\Gamma(\mathcal{R})) = \biguplus_{e \in E(Y)} \mathcal{R}(e) \backslash G \ ,$$

$$V(\Gamma(\mathcal{R})) = \biguplus_{v \in V(Y)} \mathcal{R}(v) \backslash G \ ,$$

$$(\mathcal{R}(e)g)^{\pm 1} = \mathcal{R}(e^{\pm 1})\ \mathcal{R}_{\pm 1}(e)g \ ,$$

for all $e \in E(Y)$, and all $g \in G$. One easily verifies, just as in (3.2), that the incidence maps are well-defined and that G acts on $\Gamma(\mathcal{R})$ by right-multiplication. It is also easy to check that the map $\mathcal{R}(y)g \longrightarrow y$ is a surjective map of G-graphs $\Gamma(\mathcal{R}) \longrightarrow Y$ (Y with trivial G-action), so $\Gamma(\mathcal{R}) / G$ is naturally isomorphic to Y. Identifying $\Gamma(\mathcal{R})/ G$ with Y in this way, we find that $r : \Gamma(\mathcal{R}) / G \longrightarrow \Gamma(\mathcal{R})$ with $r(y) = \mathcal{R}(y) \cdot 1$ is a transversal.

We next come to the relevant universal construction for a graph of groups $(\mathcal{Y},Y)$. Let, for each $y \in Y$, $\mathcal{U}(y) = \{(x,y) \mid x \in \mathcal{Y}(y)\}$ and set $S = \bigcup_{y \in Y} \mathcal{U}(y)$. Now define

$$\begin{aligned} \Pi(\mathcal{Y},Y) = \ <S;\ & \mathcal{U}_{\pm 1}(e),\ e \in E(Y) \mid \\ & (g,y)(h,y) = (gh,y), (g,y),(h,y) \in S\ ; \\ & \mathcal{U}_{\pm 1}(e)(g,e)\mathcal{U}_{\pm 1}(e)^{-1} = (\mathcal{Y}_{\pm 1}(e)(g), e^{\pm 1})\ , \\ & (g,e) \in S,\ e \in E(Y)> . \end{aligned}$$

These relations say nothing more than that $\Pi(\mathcal{Y},Y)$ is the most general group in which there is a realization of $(\mathcal{Y},Y)$. To state this precisely, we observe that whenever $\psi : G \longrightarrow H$ is a homomorphism of groups, there is an obvious functor $\hat{\psi} : \hat{G} \longrightarrow \hat{H}$, and that if $\mathcal{R}$ is a realization of $(\mathcal{Y},Y)$ in G, then we get a realization $\hat{\psi} \circ \mathcal{R}$ of $(\mathcal{Y},Y)$ in H. Now the presentation above for $\Pi(\mathcal{Y},Y)$ gives a canonical realization $\mathcal{U}$ of $(\mathcal{Y},Y)$ in $\Pi(\mathcal{Y},Y)$, which we call the <u>universal realization</u>, with $\mathcal{U}(y)$, $y \in Y$, and $\mathcal{U}_{\pm 1}(e)$, $e \in E(Y)$, already defined, and we have:

(4.2) <u>Let</u> $\mathcal{R}$ <u>be a realization of</u> $(\mathcal{Y},Y)$ <u>in</u> G. <u>Then there exists a unique homomorphism</u> $\psi : \Pi(\mathcal{Y},Y) \longrightarrow G$ <u>such that</u> $\mathcal{R} = \hat{\psi} \circ \mathcal{U}$. <u>In other words, every realization factors uniquely through the universal realization.</u>

<u>Proof</u>: This is an immediate consequence of the presentation for $\Pi(\mathcal{Y},Y)$. □

Now let $\Gamma(\mathcal{U})$ be the graph of the universal realization. We will prove that $\Gamma(\mathcal{U})$ has no circuits, which is really the main object of the whole exercise. The argument is essentially due to Dicks [1], but the version presented here seems a bit more transparent.

We begin by recalling some notation for semi-direct products. Thus, let G be a group and M a right G-module (a G-set with abelian group structure which is preserved by the action of G). We define multiplication on $G \ltimes M$ via:

$$(g_1,m_1)(g_2,m_2) = (g_1g_2, m_1^{g_2} + m_2) .$$

One checks that this defines a group which we denote by $G \ltimes M$ which has a normal subgroup $\{(1,m) \mid m \in M\}$ isomorphic to M and a complement $\{(g,0) \mid g \in G\}$ isomorphic to G.

Suppose that $\phi : G \longrightarrow G \ltimes M$ is a set-theoretic function of the form

$$\phi(g) = (g,d(g))$$

for some function $d : G \longrightarrow M$. One easily verifies that ϕ is a homomorphism, if and only if d satisfies

$$(4.3) \qquad d(g_1g_2) = d(g_1)^{g_2} + d(g_2) \quad \text{for all} \quad g_1,g_2 \in G\,.$$

Functions satisfying (4.3) are variously called 1-cocycles, crossed homomorphisms, or deriviations.

We can now prove:

(4.4) <u>The graph of the universal realization of any graph of groups has no circuits.</u>

<u>Proof</u>: Let $\mathcal{U}$ be the universal realization of $(\mathcal{Y},Y)$ in $\Pi = \Pi(\mathcal{Y},Y)$ and let $\Gamma = \Gamma(\mathcal{U})$. As mentioned earlier, Γ is a Π-graph, Γ/Π is naturally isomorphic to Y, and we have a transversal $u : Y \longrightarrow \Gamma$ with $u(y) = \mathcal{U}(y) \cdot 1$. Let $M = P(\Gamma)$, the space of paths in Γ, and form $\Pi \ltimes M$ as above (Π acts on M by permuting the basis $E(\Gamma)$). Now identify Π with the subgroup $\{(g,0) \mid g \in \Pi\}$ of $\Pi \times M$. Notice that if $g \in \mathcal{U}(e)$, then

$$(g,0)(1,u(e)) = (g,u(e)) = (1,u(e))(g,0)$$

since, by definition, $\mathcal{U}(e) = \Pi_{u(e)}$.

Let $\tilde{\mathcal{U}}_{+1}(e) = (\mathcal{U}_{+1}(e),u(e)) = (\mathcal{U}_{+1}(e),0)(1,u(e))$. Since $(1,u(e))$ commutes with all elements of $\mathcal{U}(e)$, ad $\mathcal{U}_{+1}(e)$ and ad $\tilde{\mathcal{U}}_{+1}(e)$ restrict to the same homomorphism on $\mathcal{U}(e)$. This means that there is a realization $\tilde{\mathcal{U}}$ of $(\mathcal{Y},Y)$ in $\Pi \ltimes M$ with $\tilde{\mathcal{U}}(y) = \mathcal{U}(y)$, $y \in Y$, $\tilde{\mathcal{U}}_{-1}(e) = (\mathcal{U}_{-1}(e),0)$ and $\tilde{\mathcal{U}}_{+1}(e)$ defined above, for $e \in E(Y)$. By (4.2) there is a homomorphism $\phi : \Pi \longrightarrow \Pi \ltimes M$ such that

$$\phi(g) = (g,0) \quad \text{for all} \quad g \in \mathcal{U}(y) \quad \text{and} \quad y \in Y ,$$

$$\left.\begin{aligned} \phi(\mathcal{U}_{-1}(e)) &= (\mathcal{U}_{-1}(e),0) \\ \phi(\mathcal{U}_{+1}(e)) &= (\mathcal{U}_{+1}(e),u(e)) \end{aligned}\right\} \quad \text{for all} \quad e \in E(Y) .$$

It follows that there exists a 1-cocycle $d : \Pi \longrightarrow M$ such that

$$d(g) = 0 \quad \text{for all} \quad g \in \bigcup_{y \in Y} \mathcal{U}(y) ,$$

$$\left.\begin{aligned} d(\mathcal{U}_{-1}(e)) &= 0 \\ d(\mathcal{U}_{+1}(e)) &= u(e) \end{aligned}\right\} \quad \text{for all} \quad e \in E(Y).$$

Now let N be the free abelian group on $V(\Gamma)$. Define $\delta : M \longrightarrow N$ via $\delta(e) = e^{+1}-e^{-1}$ for $e \in E(\Gamma)$ and extend δ linearly to M. If γ is a circuit in M, then clearly $\delta(\gamma) = 0$, so to complete the proof, we will construct a map $\eta : N \longrightarrow M$ such that $\eta(\delta(e)) = e$ for all $e \in E(\Gamma)$. Namely, let $v \in V(\Gamma)$, then $v = \mathcal{U}(y)g$ for some $g \in \Pi$, $y \in V(Y)$. Since d vanishes on $\mathcal{U}(y)$, (4.3) implies that d is constant on $\mathcal{U}(y)g$, so we put $\eta(v) = d(g)$ and extend linearly to N. Now let $e \in E(\Gamma)$, then $e = \mathcal{U}(y)g$ for some $y \in E(Y)$, and $e^{\pm 1} = \mathcal{U}(y^{\pm 1})\mathcal{U}_{\pm 1}(y)g$. It follows that

$$\begin{aligned}\eta(\delta(e)) &= \eta(e^{+1}) - \eta(e^{-1}) \\ &= d(\mathcal{U}_{+1}(y)g) - d(\mathcal{U}_{-1}(y)g) \\ &= d(\mathcal{U}_{+1}(y))^g + d(g) - d(\mathcal{U}_{-1}(y))^g - d(g) \\ &= u(y)^g = \mathcal{U}(y)g = e! \qquad \square\end{aligned}$$

The question arises whether the group $\Pi(\mathcal{Y},Y)$ "collapses" or whether the groups $\mathcal{Y}(e)$, $\mathcal{Y}(e^{\pm 1})$, $e \in E(Y)$, are faithfully represented in $\Pi(\mathcal{Y},Y)$. So, to conclude this section, we shall prove that when all $\mathcal{Y}_{\pm 1}(e)$ are monic, the universal realization is faithful. We do this by constructing a particular faithful realization in a symmetric group. To see that the condition on the $\mathcal{Y}_{\pm 1}(e)$ is necessary see (8.1) below. We first need:

(4.5) <u>Suppose</u> $\varphi : G \longrightarrow H$ <u>is an isomorphism of groups</u>. <u>Then there exists a permutation</u> σ_φ <u>of the set</u> $G \times H$ <u>such that</u> $\sigma_\varphi^2 = 1$ <u>and</u>

$$(g_1 g_2, h)^{\sigma_\varphi} = (g_1, h)^{\sigma_\varphi}(1, \varphi(g_2))$$

for all $g_1, g_2 \in G$, and all $h \in H$.

<u>Proof</u>: Choose coset representatives $\{h_i\}$ for $\varphi(G)$ in H so that every element of H is uniquely of the form $h_i\varphi(g)$ for some i and some $g \in G$. Define $(g_1, h_i\varphi(g_2))^{\sigma_\varphi} = (g_2, h_i\varphi(g_1))$ for all $g_1, g_2 \in G$ and all i. Then $\sigma_\varphi^2 = 1$ and

$$
\begin{aligned}
(g_1 g_2, h_i g)^{\sigma_\varphi} &= (g, h_i \varphi(g_1 g_2)) \\
&= (g, h_i \varphi(g_1))(1, \varphi(g_2)) \\
&= (g_1, h_i g)^{\sigma_\varphi}(1, \varphi(g_2)) \, . \qquad \square
\end{aligned}
$$

Now let $(\mathcal{Y}, Y)$ be a graph of groups with $\mathcal{Y}_{\pm 1}(e)$ monic for all $e \in E(Y)$. We shall contruct a faithful realization of $(\mathcal{Y}, Y)$ in the symmetric group $\Sigma(X)$ on the set $X = \prod_{y \in Y} \mathcal{U}(y)$. To simplify notation, whenever $B = \prod_\alpha A_\alpha$ we will identify $\Sigma(A_\alpha)$ with the subgroup of $\Sigma(B)$ fixing all factors A_β, $\beta \neq \alpha$. Now for each $y \in Y$, let $\rho_y : \mathcal{Y}(y) \longrightarrow \Sigma(\mathcal{Y}(y)) \subseteq \Sigma(X)$ be the right-regular representation: $x^{\rho_y(g)} = xg$. Let $\mathcal{R}(y)$ be the image of ρ_y, then ρ_y is an isomorphism $\mathcal{Y}(y) \simeq \mathcal{R}(y)$. For each $e \in E(Y)$ we have a pair of diagrams:

$$
\begin{array}{ccccc}
\mathcal{Y}(e) & \xrightarrow{\ \rho_e\ } & \Sigma(\mathcal{Y}(e)) & & \\
\Big\downarrow{\scriptstyle \mathcal{Y}_{\pm 1}(e)} & & & \cong & \Sigma(\mathcal{Y}(e) \times \mathcal{Y}(e^{\pm 1})) \subseteq \Sigma(X) . \\
\mathcal{Y}(e^{\pm 1}) & \xrightarrow[\ \rho_{e^{\pm 1}}\ ]{} & \Sigma(\mathcal{Y}(e^{\pm 1})) & \cong &
\end{array}
$$

If we apply (4.5) with $G = \mathcal{Y}(e)$, $H = \mathcal{Y}(e^{\pm 1})$ and $\varphi = \mathcal{Y}_{\pm 1}(e)$, we get a pair of elements $\mathcal{R}_{\pm 1}(e)$ of $\Sigma(X)$ with the property:

$$
\rho_e(g)\mathcal{R}_{\pm 1}(e) = \mathcal{R}_{\pm 1}(e)\rho_{e^{\pm 1}}(\mathcal{Y}_{\pm 1}(e)(g)) \quad \text{for all} \quad g \in \mathcal{Y}(e) \, .
$$

Then (4.1) commutes (note $\mathcal{R}_{\pm 1}(e) = \mathcal{R}_{\pm 1}(e)^{-1}$), so $(\mathcal{R},\rho)$ is a faithful realization of $(\mathcal{Y},Y)$ in $\Sigma(X)$. Then (4.2) immediately yields

(4.6) Let $(\mathcal{Y},Y)$ be a graph of groups with all $\mathcal{Y}_{\pm 1}(e)$ monic. Then the universal realization of $(\mathcal{Y},Y)$ is faithful. □

In view of (4.6) we will say that a graph of groups $(\mathcal{Y},Y)$ is faithful if all the $\mathcal{Y}_{\pm 1}(e)$ are monic. The proof of (4.6) yields the following corollary.

(4.7) A finite graph of finite groups has a faithful realization in a finite group. □

5. The fundamental group.

By a connected graph of groups we mean a graph of groups $(\mathcal{Y},Y)$ with Y connected. This is equivalent to saying that $\Gamma(\mathcal{U})$ is $\Pi(\mathcal{Y},Y)$-connected, where $\mathcal{U}$ is the universal realization of $(\mathcal{Y},Y)$. Thus, when Y is connected, $\Pi(\mathcal{Y},Y)$ acts transitively on connected components of $\Gamma(\mathcal{U})$ by (3.3). In this chapter we will assume that Y is connected.

Let $\Gamma(\mathcal{Y},Y)$ be a connected component of $\Gamma(\mathcal{U})$. By (4.4), $\Gamma(\mathcal{Y},Y)$ is a tree. We define $\pi(\mathcal{Y},Y)$ as the (set-wise) stabilizer of $\Gamma(\mathcal{Y},Y)$ in $\Pi(\mathcal{Y},Y)$. Thus $\pi(\mathcal{Y},Y)$ is well-defined up to conjugacy in $\Pi(\mathcal{Y},Y)$ and is called the fundamental group of $(\mathcal{Y},Y)$. Note that $\Gamma(\mathcal{Y},Y)/\pi(\mathcal{Y},Y) \simeq Y$, thus there is a surjective map $\tau : \Gamma(\mathcal{Y},Y) \longrightarrow Y$. A connected realization of $(\mathcal{Y},Y)$ is a realization $\mathcal{R}$ with $\Gamma(\mathcal{R})$ connected.

(5.1) Suppose that $\mathcal{R}$ is a connected realization of $(\mathcal{Y},Y)$ in G. Then $G = \langle \mathcal{R}(y), \mathcal{R}_{\pm 1}(e) \mid y \in Y,\ e \in E(Y)\rangle$.

Proof: Since $\Gamma(\mathcal{R})/G$ is naturally isomorphic to Y, the map $r : Y \longrightarrow \Gamma(\mathcal{R})$ with $r(y) = \mathcal{R}(y)\cdot 1$ is a transversal, which is augmented by the elements $\mathcal{R}_{\pm 1}(e)$ of G, $e \in E(Y)$. Since $\mathcal{R}(y) = G_{r(y)}$ for all $y \in Y$, the result follows from (3.6). □

(5.2) Suppose that $\mathcal{R}$ is a connected realization of $(\mathcal{Y},Y)$ in G, and let $\phi : \Pi(\mathcal{Y},Y) \longrightarrow G$ be the factorization of $\mathcal{R}$ through $\mathcal{U}$ as in (4.2). Then $\phi_\pi = \phi|_{\pi(\mathcal{Y},Y)} : \pi(\mathcal{Y},Y) \longrightarrow G$ is surjective and if $N_\pi = \ker \phi_\pi$, there is a natural isomorphism $\Gamma(\mathcal{Y},Y)/N_\pi \simeq \Gamma(\mathcal{R})$.

Proof: By (5.1) $\phi : \Pi(\mathcal{Y},Y) \longrightarrow G$ is surjective. Let $N = \ker \phi$, then it follows from the definitions that there is a natural isomorphism $\Gamma(\mathcal{R}) \simeq \Gamma(\mathcal{U})/N$. Namely, ϕ induces a surjective mapping $\phi^* : \Gamma(\mathcal{U}) \longrightarrow \Gamma(\mathcal{R})$ given by $\phi^*(\mathcal{U}(y)g) = \mathcal{R}(y)\phi(g)$, $y \in Y$, and the fibers A_β of ϕ^*, where $A_\beta = \{\alpha \in \Gamma(\mathcal{U}) \mid \phi^*(\alpha) = \beta\}$, $\beta \in \Gamma(\mathcal{R})$, are easily seen to be the N-orbits of $\Gamma(\mathcal{U})$. Now by (3.3), $\phi^*|_{\Gamma(\mathcal{Y},Y)}$ is surjective and therefore induces a natural isomorphism $\Gamma(\mathcal{Y},Y)/N_\pi \simeq \Gamma(\mathcal{R})$ where $N_\pi = N \cap \pi(\mathcal{Y},Y)$. Since $\Gamma(\mathcal{U})/N$ is connected, N is transitive on connected components of $\Gamma(\mathcal{U})$, and thus $\Pi(\mathcal{Y},Y) = \pi(\mathcal{Y},Y)N$ by definition of the fundamental group. We conlude that $\phi_\pi : \pi(\mathcal{Y},Y) \longrightarrow G$ is surjective. □

As a corollary to (5.2) we obtain:

(5.3) Let Γ be a connected G-graph with $Y = \Gamma/G$. Let $t : Y \longrightarrow \Gamma$ be an augmented transversal and let $(\mathcal{Y},Y)$ be the resulting graph of groups with realization $\mathcal{R}$ in G. Then there is a normal subgroup N of $\pi(\mathcal{Y},Y)$ acting freely on $\Gamma(\mathcal{Y},Y)$ and isomorphisms $G \simeq \pi(\mathcal{Y},Y)/N$, $\Gamma \simeq \Gamma(\mathcal{Y},Y)/N$. In particular, if Γ is a tree then $G \simeq \pi(\mathcal{Y},Y)$.

Proof: By (3.2) $\Gamma \simeq \Gamma(\mathcal{R})$ so $\mathcal{R}$ is a connected realization of $(\mathcal{Y},Y)$. Let $\phi : \Pi(\mathcal{Y},Y) \longrightarrow G$ be the factorization of $\mathcal{R}$ through the universal realization $\mathcal{U}$, then ϕ is a local isomorphism, since in this case $\mathcal{Y}(y) \longrightarrow \mathcal{U}(y) \xrightarrow{\phi} \mathcal{R}(y) = \mathcal{Y}(y)$ is the identity map. In particular, N_π acts freely on $\Gamma(\mathcal{Y},Y)$ and the result follows from (5.2) and (3.4). □

(5.4) Suppose G acts freely on a tree Γ. Then $G \simeq \pi(\mathit{Id},\Gamma/G)$.

Proof: Here $\mathcal{I}d$ means the trivial functor $\mathcal{I}d(y) = 1$ for all $y \in \Gamma/G$. Since $G_{t(y)} = 1$ for all $y \in \Gamma/G$, we simply apply (5.3). □

(5.5) The normal subgroup N of (5.3) is isomorphic to $\pi(\mathcal{I}d,\Gamma)$.

Proof: This follows from (5.4) since N acts freely on $\Gamma(\mathcal{Y},Y)$. □

6. Generators and relations for $\pi(\mathcal{Y},Y)$.

In this chapter $(\mathcal{Y},Y)$ is a connected graph of groups. We now obtain generators and relations for $\pi(\mathcal{Y},Y)$. The idea is to choose a "good" transversal $t : Y \longrightarrow \Gamma(\mathcal{Y},Y)$. Namely, we want to have $t(e^{\pm 1}) = t(e)^{\pm 1}$ for a large subset of edges e of Y. This enables us to take $t_{\pm 1}(e) = 1$ in (3.1). Thus we consider "partial splittings" for the natural map $\tau : \Gamma(\mathcal{Y},Y) \longrightarrow Y$, that is, we let Σ be the set of pairs (t,T) where T is a subgraph of Y, $t : T \longrightarrow \Gamma(\mathcal{Y},Y)$ is a graph morphism, and $\tau \cdot t = 1_T$. In particular, $\tau \cdot t$ is injective and so T has no circuits. Σ is partially ordered by inclusion-extension: $(t,T) << (t',T')$ if $T \subseteq T'$ and $t'|_T = t$. Now, let T_o be any subtree of Y and Σ_{T_o} be those elements (t,T) of Σ with $T \subseteq T_o$. Σ_{T_o} is obviously non-empty since we can take $T = \{y\}$ for any $y \in V(T_o)$. By Zorn's lemma Σ_{T_o} has a maximal element (t,T). Suppose $T \subset T_o$. Then there exists an edge $e \in E(T_o)-E(T)$ with $e^{+1} \in V(T)$, $e^{-1} \notin V(T)$ (or $e^{+1} \notin V(T)$, $e^{-1} \in V(T)$). By local surjectivity we find an edge $f \in E(\Gamma(\mathcal{Y},Y))$ with $\tau(f) = e$ and $\tau(f^{-1}) = \tau(f)^{-1}$, $\tau(f^{+1}) = \tau(f)^{+1}$. Define a subtree T' of T_o with $V(T') = V(T) \cup \{e^{-1}\}$ (resp. $V(T') = V(T) \cup \{e^{+1}\}$) and $E(T') = E(T) \cup \{e\}$. We get a graph morphism $t' : T' \longrightarrow \Gamma(\mathcal{Y},Y)$ by setting $t'|_T = t$ and $t'(e^{-1}) = f^{-1}$ (resp. $t'(e^{+1}) = f^{+1}$). It follows $(t,T) << (t',T')$ and maximality of (t,T) gives $T = T_o$. Now another application of Zorn's lemma produces a maximal element (t,T) of Σ with $T_o \subseteq T$ and the same argument as above gives $V(T) = V(Y)$.

How to extend t to the remainder of $E(Y)$? By local surjectivity we can choose $t(e) \in \Gamma(\mathcal{Y},Y)$ with $t(e^{-1}) = t(e)^{-1}$ for $e \in E(Y)-E(T)$. Thus we have proved:

(6.1) Let T_0 be any subtree of Y. Then there exists a subtree T of Y and a transversal $t : Y \longrightarrow \Gamma(\mathcal{Y},Y)$ such that:

(a) $T_0 \subseteq T$,

(b) $V(T) = V(Y)$,

(c) $t(e^{+1}) = t(e)^{+1}$ for all $e \in E(T)$,

(d) $t(e^{-1}) = t(e)^{-1}$ for all $e \in E(Y)$. □

Now take a transversal t satisfying (6.1), and augment it with elements $t_{\pm 1}(e) \in \pi(\mathcal{Y},Y)$ subject to (3.1). Because of conditions (c) and (d) of (6.1), however, we may, and do, take $t_{-1}(e) = 1$ for all $e \in E(Y)$, and $t_{+1}(e) = 1$ for all $e \in E(T)$. Let $t(y) = \mathcal{U}(y)g_y$, and let $\mathcal{R}$ be the resulting realization of $(\mathcal{Y},Y)$ in $\pi(\mathcal{Y},Y)$ (with $\mathcal{R}(y) = \Pi_{t(y)}$, where $\Pi = \Pi(\mathcal{Y},Y)$, and with $\rho_y : \mathcal{Y}(y) \longrightarrow \mathcal{R}(y)$ the composite $\mathcal{Y}(y) \longrightarrow \mathcal{U}(y) \xrightarrow{\text{ad } g_y^{-1}} \mathcal{R}(y)$, and $\mathcal{R}_{\pm 1}(e) = t_{\pm 1}(e)$). Then $\mathcal{R}_{-1}(e) = 1$ for all $e \in E(Y)$ and $\mathcal{R}_{+1}(e) = 1$ for all $e \in E(T)$. We shall say that a realization which satisfies these conditions is reduced at T.

(6.2) Let T_0 be a subtree of Y. Then there exists a maximal subtree T of Y containing T_0 and a realization $\mathcal{R}$ of $(\mathcal{Y},Y)$ in $\pi(\mathcal{Y},Y)$ which is reduced at T. Let $\phi : \Pi(\mathcal{Y},Y) \longrightarrow \pi(\mathcal{Y},Y)$ with $\mathcal{R} = \hat{\phi} \circ \mathcal{U}$. Then $\ker \phi$ is the smallest normal subgroup of $\Pi(\mathcal{Y},Y)$ containing $\{\mathcal{U}_{-1}(e) \mid e \in E(Y)\} \cup \{\mathcal{U}_{+1}(e) \mid e \in E(T)\}$.

Proof: The existence of a realization $\mathcal{R}$ reduced at T was established above. Let N be the smallest normal subgroup of $\Pi = \Pi(\mathcal{Y},Y)$ containing $\{\mathcal{U}_{-1}(e) \mid e \in E(Y)\} \cup \{\mathcal{U}_{+1}(e) \mid e \in E(T)\}$, then $N \subseteq \ker \phi$. Note that ϕ is a local isomorphism because $\phi|_{\mathcal{U}(y)} = \mathrm{ad}\ g_y^{-1}$.

Also note that $\Gamma(\mathcal{R}) \simeq \Gamma(\mathcal{Y},Y)$ by (3.2). Let $\bar{\Gamma} = \Gamma(\mathcal{U})/N$, then since $N \subseteq \ker \phi$, ϕ induces a local isomorphism of $\bar{\Gamma}$ onto $\Gamma(\mathcal{Y},Y)$. To show that $N = \ker \phi$ it suffices, by (3.4), to show that $\bar{\Gamma}$ is connected. Let $u : Y \longrightarrow \Gamma(\mathcal{U})$ be the transversal $u(y) = \mathcal{U}(y) \cdot 1$, $y \in Y$. Then $\bar{u}(y) = \overline{u(y)} = \mathcal{U}(y) \cdot N$ is a transversal $Y \longrightarrow \bar{\Gamma}$. Since N contains $\mathcal{U}_{-1}(e)$, $e \in E(Y)$, and $\mathcal{U}_{+1}(e)$, $e \in E(T)$, it follows from the definition of $\Gamma(\mathcal{U})$ that

$$\bar{u}(e^{-1}) = \bar{u}(e)^{-1}, \quad e \in E(Y),$$
$$\bar{u}(e^{+1}) = \bar{u}(e)^{+1}, \quad e \in E(T).$$

These relations imply that $\{\bar{u}(y) \mid y \in Y\}$ is contained in a single connected component $\bar{\Gamma}_o$ of $\bar{\Gamma}$. But $\bar{\Gamma}$ is an Π-connected Π-graph, and $\Pi_{\bar{u}(y)} = \mathcal{U}(y)$, since $\bar{u}(y)$ lies in $\bar{\Gamma}_o$, $\mathcal{U}(y)$ and $\mathcal{U}_{\pm 1}(e)$ stabilize $\bar{\Gamma}_o$ for all y and e. Since $\Pi = \langle \mathcal{U}(y), \mathcal{U}_{\pm 1}(e) \mid y \in Y, e \in E(Y)\rangle$ we get $\bar{\Gamma}_o = \bar{\Gamma}$. □

The above result says that a presentation for $\pi(\mathcal{Y},Y)$ is obtained by adding to the defining relations for $\Pi(\mathcal{Y},Y)$ the relations $\mathcal{U}_{-1}(e) = 1$, $e \in E(Y)$, $\mathcal{U}_{+1}(e) = 1$, $e \in E(T)$. From this we can, for example, find an explicit set of the generators for a group acting freely on a tree.

(6.3) <u>Suppose that</u> G <u>acts freely on a tree</u> Γ, <u>and let</u> $\bar{\Gamma} = \Gamma/G$. <u>Then there exists a subtree</u> Γ_o <u>of</u> Γ <u>such that</u> $\bar{\Gamma}_o$ <u>is a maximal subtree of</u> $\bar{\Gamma}$. <u>For any such</u> Γ_o, <u>and for each edge</u> $\bar{e} \in E(\bar{\Gamma})-E(\bar{\Gamma}_o)$, <u>choose a representative</u> e <u>with</u> $e^{-1} \in V(\Gamma_o)$, <u>and an element</u> $g_{\bar{e}} \in G$ <u>with</u> $(e^{+1})^{g_{\bar{e}}} \in V(\Gamma_o)$.

Then $\{g_{\bar e} \mid \bar e \in E(\bar\Gamma)-E(\bar\Gamma_o)\}$ *is a set of free generators for* G.

Proof: By (5.4) we have $G \simeq \pi(\mathcal{I}d,\bar\Gamma)$. Since $\Pi(\mathcal{I}d,\bar\Gamma)$ is the free group on generators $\{\mathcal{U}_{\pm 1}(e) \mid e \in E(\bar\Gamma)\}$ by definition, the result follows from (6.2). □

As a further useful result we have the following:

(6.4) *Let* $\mathcal{R}$ *be the realization of* $(\mathcal{Y},Y)$ *in* $\pi(\mathcal{Y},Y)$ *given in* (6.2). *Set* $\theta(\mathcal{Y},Y)$ *to be the smallest normal subgroup of* $\pi(\mathcal{Y},Y)$ *generated by all the vertex-groups* $\mathcal{R}(v)$, $v \in V(Y)$. *Then* $\pi(\mathcal{Y},Y)/\theta(\mathcal{Y},Y)$ *is isomorphic to* $\pi(\mathcal{I}d,Y)$.

Proof: The presentation of (6.2) shows that $\pi(\mathcal{Y},Y)/\theta(\mathcal{Y},Y)$ is defined by the generators $\mathcal{U}_{+1}(e)$, $e \in E(Y)\setminus E(T)$. The result now follows from (6.3). □

7. The fundamental group - revisited; normal forms.

Our goal in this section is to present a second description of the fundamental group of a connected faithful graph of groups $(\mathcal{Y},Y)$. This description has the advantage that the elements of $\pi(\mathcal{Y},Y)$ can be brought into a particular useful canonical form. Throughout we will assume that $(\mathcal{Y},Y)$ is a connected faithful graph of groups, $t : Y \longrightarrow \Gamma(\mathcal{Y},Y)$ is a transversal satisfying (6.1) with respect to the subtree T of Y, and $\mathcal{U}$ is the universal realization of $(\mathcal{Y},Y)$ in $\Pi(\mathcal{Y},Y)$.

To begin with, note that the definition of a path $\gamma = \sum_{i=1}^{n} \varepsilon_i e_i$ does not permit $e_i = e_{i+1}$. In order to allow for such "backtracking" we make the following general definitions. Let Γ be any graph, set $W(\Gamma)$ to be the set of all finite sequences of the form

$$(v_o,\varepsilon_1 e_1,v_1,\ldots, v_{i-1},\varepsilon_i e_i, v_i,\ldots, v_{n-1},\varepsilon_n e_n,v_n)$$

where $v_i \in V(\Gamma)$, $e_i \in E(\Gamma)$, $\varepsilon_i = \pm 1$ and

$$v_i = e_{i+1}^{-\varepsilon_{i+1}} \quad , \quad 0 \le i \le n-1 \; ,$$

$$v_i = e_i^{\varepsilon_i} \quad , \quad 1 \le i \le n \; .$$

An element of $W(\Gamma)$ is called a walk and can be viewed as a path γ together with instructions on how to "transverse" the path. Note that when $n = o$ a walk is simply a sequence (v_o) where $v_o \in V(\Gamma)$.

We now return to our concrete situation with $(\mathcal{Y},Y)$ a connected, faithful graph of groups. For any edge e of Y we define the element Q_e of $\Pi(\mathcal{Y},Y)$ by

$$Q_e = \mathcal{U}_{+1}(e)\mathcal{U}_{-1}(e)^{-1} .$$

Let σ be a walk in Y. A σ-<u>sequence</u> is an $(n+1)$-tupel $\tau = (g_0,\ldots,g_n)$ of elements $g_i \in \mathcal{U}(v_i)$, $0 \leq i \leq n$. To each walk σ and σ-sequence τ there is an associated σ-<u>word</u> (σ,τ) in $\Pi(\mathcal{Y},Y)$ given by

$$(\sigma,\tau) = g_n\, Q_{e_n}^{\varepsilon_n}\, g_{n-1}\, Q_{e_{n-1}}^{\varepsilon_{n-1}} \cdots g_1\, Q_{e_1}^{\varepsilon_1}\, g_0 .$$

(Note that $Q_{e_i}^{\varepsilon_i} = \mathcal{U}_{\varepsilon_i}(e_i)\mathcal{U}_{-\varepsilon_i}(e_i)^{-1}$.) For any pair of (not necessarily distinct) vertices v,w of Y there is a unique path $\gamma(v,w) = \sum_{i=1}^{n} \varepsilon_i e_i$ with $e_i \in E(T)$ and $v = e_1^{-\varepsilon_1}$, $w = e_n^{\varepsilon_n}$ (if $v = w$, then $\gamma(v,v) = 0$). If $v \neq w$, set $\lambda(v,w) = \prod_{i=1}^{n} Q_{e_i}^{-\varepsilon_i}$, and if $v = w$, set $\lambda(v,w) = 1$.

Choose a vertex v in Y and let $\pi(\mathcal{Y},Y;\, v)$ be the set of all σ-words (σ,τ) in $\Pi(\mathcal{Y},Y)$ where $\sigma = (v_0,\varepsilon_1 e_1,\ldots,\varepsilon_n e_n,v_n)$ is any walk satisfying $v_0 = v = v_n$. It is immediate that $\pi(\mathcal{Y},Y;\, v)$ is a subgroup of $\Pi(\mathcal{Y},Y)$, the so-called <u>fundamental group of</u> $(\mathcal{Y},Y)$ <u>at the vertex</u> v. We now define for $w \in V(\Gamma)$, $e \in E(\Gamma)$:

$$(7.1)\qquad \begin{aligned} \mathcal{T}(w) &= \lambda(v,w)\ \mathcal{U}(w)\ \lambda(v,w)^{-1} \\ \mathcal{T}(e) &= \lambda(v,e^{-1})\ \mathcal{U}_{-1}(e)\ \mathcal{U}(e)\ \mathcal{U}_{-1}(e)^{-1}\ \lambda(v,e^{-1})^{-1} \\ \mathcal{T}_{-1}(e) &= 1 \\ \mathcal{T}_{+1}(e) &= \lambda(v,e^{+1})\ Q_e\ \lambda(v,e^{-1})^{-1} \end{aligned}$$

It is easy to check that (7.1) defines a realization of $(\mathcal{Y},Y)$ in $\pi(\mathcal{Y},Y;\ v)$. Thus (4.2) produces a unique homomorphism $\psi : \Pi(\mathcal{Y},Y) \longrightarrow \pi(\mathcal{Y},Y;\ v)$.

Let $e \in E(T)$. It follows immediately from the definitions that $\mathcal{T}_{\pm 1}(e) = 1$. Since we already have that $\mathcal{T}_{-1}(e) = 1$ for all $e \in E(Y)$, it follows from (6.2) that ψ factors through $\pi(\mathcal{Y},Y)$; that is, there is a unique homomorphism $\tilde{\psi} : \pi(\mathcal{Y},Y) \longrightarrow \pi(\mathcal{Y},Y;\ v)$ making the following diagram:

$$(7.2)\qquad \begin{array}{ccc} \Pi(\mathcal{Y},Y) & \xrightarrow{\ \psi\ } & \pi(\mathcal{Y},Y;\ v) \\ \phi \downarrow & \nearrow_{\tilde{\psi}} & \\ \pi(\mathcal{Y},Y) & & \end{array}$$

commute. Furthermore, $\phi(\lambda(v,w)) = 1$ for each $w \in V(Y)$, so letting $i : \pi(\mathcal{Y},Y;\ v) \longrightarrow \Pi(\mathcal{Y},Y)$ be the canonical injection we have $\phi \circ i \circ \tilde{\psi} = id$.

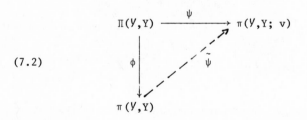

On the other hand, we now show that ψ is a retraction of $\Pi(\mathcal{Y},Y)$ onto $\pi(\mathcal{Y},Y;\ v)$; that is, $\psi \circ i = id$, which will show that $\tilde{\psi}$ is an isomorphism. Let $\sigma = (v_0,\varepsilon_1 e_1,\ldots,\varepsilon_n e_n,v_n)$ be a walk in Y with $v_0 = v = v_n$, $\tau = (g_0,\ldots,g_n)$ a σ-sequence and (σ,τ) the associated σ-word. The commutativity of (7.2) together with the definition of the realization T shows that

$$\psi(g_i) = \lambda(v,v_i)\, g_i\, \lambda(v,v_i)^{-1} \quad , \quad 0 \le i \le n$$
$$\psi(Q_e) = \lambda(v,e^{+1})\, Q_e\, \lambda(v,e^{-1})^{-1} \quad , \quad e \in E(Y) .$$

For $1 \le i \le n$ we have

$$\psi(g_i Q_i^{\varepsilon_i}) = \lambda(v,v_i)\, g_i\, \lambda(v,v_i)^{-1}\, [\lambda(v,e_i^{+1})\, Q_{e_i}\, \lambda(v,e_i^{-1})^{-1}]^{\varepsilon_i} .$$

One checks easily that for both $\varepsilon_i = +1$ as well as for $\varepsilon_i = -1$ this gives

$$\psi(g_i Q_i^{\varepsilon_i}) = \lambda(v,v_i)\, g_i\, Q_{e_i}^{\varepsilon_i}\, \lambda(v,v_{i-1})^{-1} .$$

Hence

$$\psi((\sigma,\tau)) = \left(\prod_{i=n}^{1} \lambda(v,v_i)\, g_i\, Q_{e_i}^{\varepsilon_i}\, \lambda(v,v_{i-1})^{-1}\right) g_0$$

which collapses to give

$$\psi((\sigma,\tau)) = (\sigma,\tau) .$$

Hence ψ is a retraction and we conclude that $id = \psi \circ i = \tilde{\psi} \circ \phi \circ i$. We have proven the following:

(7.3) The canonical homomorphism $\phi : \Pi(\mathcal{Y},Y) \longrightarrow \pi(\mathcal{Y},Y)$ induces an isomorphism of $\pi(\mathcal{Y},Y;\, v)$ onto $\pi(\mathcal{Y},Y)$. In particular, the isomorphism type of $\pi(\mathcal{Y},Y;\, v)$ is independent of the choice of v. □

Let $g \in \Pi(\mathcal{Y},Y)$ with $g = (\sigma,\tau) = (\sigma',\tau')$. It is not necessarily true that $\sigma = \sigma'$, $\tau = \tau'$. We will describe two ways in which an element can have different representations of this type. We will later see that these two ways essentially exhaust all the possibilities.

Suppose first that $\sigma = (v_0,\varepsilon_1 e_1,\ldots,\varepsilon_n e_n,v_n)$ is a walk in Y, $\tau = (g_0,\ldots,g_n)$ is a σ-sequence and (σ,τ) the associated σ-word. If for some i we have $e_i = e_{i+1}$, $\varepsilon_i = -\varepsilon_{i+1}$, and $g_i \in \mathcal{U}_{\varepsilon_i}(e_i)\mathcal{U}(e_i)\mathcal{U}_{\varepsilon_i}(e_i)^{-1}$, then $Q_{e_{i+1}}^{\varepsilon_{i+1}}\, g_i\, Q_{e_i}^{\varepsilon_i} \in \mathcal{U}(v_{i-1}) = \mathcal{U}(v_{i+1})$. This means that $(\sigma,\tau) = (\sigma',\tau')$ where

$$\sigma' = (v_0,\ldots,v_{i-2},\varepsilon_{i-1}e_{i-1},v_{i-1},\varepsilon_{i+2}e_{i+2},v_{i+2},\ldots,v_n)$$

and $$\tau' = (g_0,\ldots,g_{i-2},g',g_{i+2},\ldots,g_n)\ , \text{ where}$$

$$g' = g_{i+1}\, Q_{e_{i+1}}^{\varepsilon_{i+1}}\, g_i\, Q_{e_i}^{\varepsilon_i}\, g_{i-1} \in \mathcal{U}(v_{i-1}) = \mathcal{U}(v_{i+1})\ .$$

We say that (σ',τ') is a reduction of (σ,τ) (at e_i, if we wish to be specific). Evidently, to any σ-word we can apply a sequence of reductions until we reach a point where no further reduction is possible. Such a word is then said to be reduced.

The second way in which σ-words can be equal is a little more subtle. We require the following definition. Assume that (σ,τ) and $(\sigma\ ,\tau')$

are two σ-words with $\tau' = (g'_o,\ldots,g'_n)$. We say that τ and τ' are σ-equivalent if there is a sequence $(a_1,\ldots,a_n)$ with $a_i \in \mathcal{U}(e_i)$ such that

$$g_n = g'_n\, \mathcal{U}_{\varepsilon_n}(e_n) a_n\, \mathcal{U}_{\varepsilon_n}(e_n)^{-1},$$

$$\mathcal{U}_{-\varepsilon_{i+1}}(e_{i+1}) a_{i+1}\, \mathcal{U}_{-\varepsilon_{i+1}}(e_{i+1})^{-1} g_i = g'_i\, \mathcal{U}_{\varepsilon_i}(e_i) a_i\, \mathcal{U}_{\varepsilon_i}(e_i)^{-1}, \qquad 1 \le i < n,$$

$$\mathcal{U}_{-\varepsilon_1}(e_i) a_1\, \mathcal{U}_{-\varepsilon_1}(e_1)^{-1} g_o = g'_o .$$

Note that if $n = o$, then τ and τ' are σ-equivalent if and only if $g_o = g'_o$.

(7.4) _Suppose_ τ _and_ τ' _are_ σ-_equivalent, then_ $(\sigma,\tau) = (\sigma,\tau')$. _Moreover, if_ (σ,τ) _is reduced, so is_ (σ,τ').

Proof:

$$(\sigma,\tau) = g_n\, Q_{e_n}^{\varepsilon_n}\, g_{n-1}\, Q_{e_{n-1}}^{\varepsilon_{n-1}} \cdots g_1\, Q_{e_1}^{\varepsilon_1}\, g_o$$

$$= g'_n\, \mathcal{U}_{\varepsilon_n}(e_n) a_n\, \mathcal{U}_{\varepsilon_n}(e_n)^{-1}\, Q_{e_n}^{\varepsilon_n}\, g_{n-1} \cdots g_o$$

$$= g'_n\, Q_{e_n}^{\varepsilon_n}\, \mathcal{U}_{-\varepsilon_n}(e_n) a_n\, \mathcal{U}_{-\varepsilon_n}(e_n)^{-1}\, g_{n-1} \cdots g_o$$

$$= g'_n\, Q_{e_n}^{\varepsilon_n}\, g'_{n-1}\, \mathcal{U}_{\varepsilon_{n-1}}(e_{n-1}) a_{n-1}\, \mathcal{U}_{\varepsilon_{n-1}}(e_{n-1})^{-1}\, Q_{e_{n-1}}^{\varepsilon_{n-1}} \cdots g_o$$

$$= \ldots$$

$$= g'_n\, Q_{e_n}^{\varepsilon_n}\, g'_{n-1}\, Q_{e_n}^{\varepsilon_{n-1}} \cdots g'_1\, Q_{e_1}^{\varepsilon_1}\, g'_o$$

$$= (\sigma,\tau').$$

Clearly, if (σ,τ) is reduced, so is (σ,τ'). □

In order to show that reduction and σ-equivalence account for all the possible ways in which two σ-words may be equal we first need two preliminary results. The first one has a long history and in the case that $Y =$  is known as "Britton's Lemma". It has had considerable application in proving the unsolvability of several well-known problems - in particular, the word problem for groups.

(7.5) Let v be a vertex in Y. Then every element of $\mathcal{U}(v)$ can be uniquely expressed as a reduced σ-word (σ,τ). This unique expression requires $\sigma = (v)$.

Proof: Let $g \in \mathcal{U}(v)$. Since $\mathcal{U}(v)$ is a subgroup of $\pi(\mathcal{Y},Y;\, v)$ we can assume to the contrary that g can be expressed as a reduced σ-word $g = (\sigma,\tau)$ with $\sigma = (v_0,\varepsilon_1 e_1,\ldots,\varepsilon_n e_n,v_n)$ and $n \geq 1$. For $1 \leq i \leq n$ set $\sigma_i = (v_0,\varepsilon_1 e_1,\ldots,\varepsilon_{i-1}e_{i-1},v_{i-1})$, $\tau_i = (g_0,\ldots,g_{i-1})$, and $\eta_i = \mathcal{U}(e_i)\,\mathcal{U}_{\varepsilon_i}(e_i)^{-1}\,Q_{e_i}^{\varepsilon_i}(\sigma_i,\tau_i)$. Let η be the element of $W(\Gamma(\mathcal{Y},Y))$ given by

$$\eta = (\eta_1^{-\varepsilon_1},\varepsilon_1\eta_1,\eta_2^{-\varepsilon_2},\ldots,\eta_n^{-\varepsilon_n},\varepsilon_n\eta_n,\eta_n^{\varepsilon_n}).$$

Using the definition of the adjacency maps in the graph of a realization, one checks that

$$\eta_i^{-\varepsilon_i} = \mathcal{U}(e_i^{-\varepsilon_i})\, g_{i-1}\, Q_{e_{i-1}}^{\varepsilon_{i-1}} \cdots g_o \,,$$

$$\eta_i^{\varepsilon_i} = \mathcal{U}(e_i^{\varepsilon_i})\, Q_{e_i}^{\varepsilon_i}\, g_{i-1}\, Q_{e_{i-1}}^{\varepsilon_{i-1}} \cdots g_o \,,$$

$$= \mathcal{U}(e_i^{\varepsilon_i})\, g_i\, Q_{e_i}^{\varepsilon_i}\, g_{i-1}\, Q_{e_{i-1}}^{\varepsilon_{i-1}} \cdots g_o \qquad (\text{as } g_i \in \mathcal{U}(e_i^{\varepsilon_i})),$$

in particular,

$$\eta_n^{\varepsilon_n} = \mathcal{U}(v) = \eta_1^{-\varepsilon_1} \qquad (\text{as } g \in \mathcal{U}(v)).$$

Set $\gamma = \sum_{i=1}^{n} \varepsilon_i \eta_i$, so γ is a circuit in $\Gamma(\mathcal{Y},Y)$. But $\Gamma(\mathcal{Y},Y)$ has no circuits, hence $\gamma = 0$ and there must be some edge η_i followed by its negative. On examining this edge we find that $e_i = e_{i+1}$ in Y and $g_i \in \mathcal{U}_{\varepsilon_i}(e_i)\, \mathcal{U}(e_i)\, \mathcal{U}_{\varepsilon_i}(e_i)^{-1}$. But this contradicts the fact that (σ,τ) is reduced. □

(7.6) <u>Suppose that</u> (σ,τ) <u>and</u> (σ,τ') <u>are reduced and</u> $(\sigma,\tau) = (\sigma,\tau')$. <u>Then</u> τ <u>and</u> τ' <u>are</u> σ-<u>equivalent</u>.

<u>Proof</u>: If $n = 0$, this is obvious. Suppose then that $n > 0$, $\tau = (g_o,\ldots,g_n)$, $\tau' = (g'_o,\ldots,g'_n)$. We proceed by induction on n. Set

$$\rho = (v_o,\varepsilon_1 e_1,\ldots,v_{n-1},\varepsilon_n e_n,v_n,-\varepsilon_n e_n,v_{n-1},\ldots,-\varepsilon_1 e_1,v_o)$$
$$\nu = (g_o,g_1,\ldots,(g'_n)^{-1} g_n,(g'_{n-1})^{-1},\ldots,(g'_o)^{-1}).$$

Evidently, ρ is a walk in Γ, ν is a ρ-sequence and $(\rho,\nu) = 1$. By (7.5) applied to the element $1 \in \mathcal{U}(e_1^{-\varepsilon_1})$ we see that (ρ,ν) is not reduced. Let (ρ',ν') be a reduction of (ρ,ν), say at e_i. If $i \neq n$, then either (σ,τ) or (σ,τ') is not reduced. So $i = n$ and $(g_n')^{-1} g_n \in \mathcal{U}_{\varepsilon_n}(e_n)\, \mathcal{U}(e_n)\, \mathcal{U}_{\varepsilon_n}(e_n)^{-1}$. So there exists some $a_n \in \mathcal{U}(e_n)$ with

$$g_n = g_n' \,\mathcal{U}_{\varepsilon_n}(e_n) a_n \,\mathcal{U}_{\varepsilon_n}(e_n)^{-1} .$$

Set $\sigma_o = (v_o, \varepsilon_1 e_1, \ldots, v_{n-2}, \varepsilon_{n-1} e_{n-1}, v_{n-1})$, $\tau_o = (g_o, \ldots, g_{n-2}, \mathcal{U}_{-\varepsilon_n}(e_n) a_n \,\mathcal{U}_{-\varepsilon_n}(e_n)^{-1} g_{n-1})$, $\tau_o' = (g_o', \ldots, g_{n-1}')$. One checks that $(\sigma_o,\tau_o) = (\sigma_o,\tau_o')$ and, moreover, (σ_o,τ_o) and (σ_o,τ_o') are reduced. By the induction hypothesis τ_o and τ_o' are σ_o-equivalent. This immediately yields that τ and τ' are σ-equivalent. The result holds. □

We can now make precise the result about uniqueness of representation.

(7.7) <u>Let</u> $v \in V(Y)$, $g \in \pi(\mathcal{Y},Y;\, v)$. <u>There is a walk</u> σ, <u>a</u> σ-<u>sequence</u> τ, <u>and a reduced</u> σ-<u>word</u> (σ,τ) <u>so that</u> $g = (\sigma,\tau)$. <u>Moreover, these conditions determine</u> σ <u>uniquely and</u> τ <u>up to equivalence</u>.

<u>Proof</u>: The existence of (σ,τ) with the stated properties follows from the definition of $\pi(\mathcal{Y},Y;\, v)$ and that of reduction. If $g \in \mathcal{U}(v)$, then the uniqueness is (7.5). Hence we may assume that $\sigma = (v_o, \varepsilon_1 e_1, \ldots, \varepsilon_n e_n, v_n)$, $v_o = v = v_n$, and $n \geq 1$. Suppose further that $g = (\sigma',\tau')$ with $\sigma' = (w_o, \lambda_1 f_1, \ldots, \lambda_m f_m, w_m)$, $\tau' = (g_o', \ldots, g_m')$. Once we know that $\sigma = \sigma'$, (7.6) will give the full result.

The proof is now similar to those of (7.5) and (7.6). Set

$$\rho = (v_o,\varepsilon_1 e_1,\ldots,\varepsilon_n e_n, v_n, -\lambda_m f_m,\ldots,-\lambda_1 f_1, v_o)$$

$$= (v_o,\varepsilon_1, e_1,\ldots,\varepsilon_n e_n, v_n, \varepsilon_{n+1} e_{n+1},\ldots,\varepsilon_{n+m} e_{n+m}, v_{n+m}),$$

$$\nu = (g_o,\ldots,g_{n-1},(g'_m)^{-1} g_n,(g'_{m-1})^{-1},\ldots,(g'_o)^{-1})$$

$$= (h_o,\ldots,h_{n-1},h_n,h_{n+1},\ldots,h_{n+m}) .$$

Evidently, ρ is a walk, ν is a ρ-sequence, and $(\rho,\nu) = 1$. For $1 \leq i \leq n+m$ set $\sigma_i = (v_o,\varepsilon_1 e_1,\ldots,\varepsilon_{i-1} e_{i-1}, v_{i-1})$, $\tau_i = (h_o,\ldots,h_{i-1})$, $\eta_i = \mathcal{U}(e_i)\, \mathcal{U}_{\varepsilon_i}(e_i)^{-1}\, Q_{e_i}^{\varepsilon_i}(\sigma_i,\tau_i)$. As in (7.5) one checks that $\eta_{n+m}^{\varepsilon_{n+m}} = \mathcal{U}(v) = \eta_1^{-\varepsilon_1}$. Set $\gamma = \sum_{i=1}^{n+m} \varepsilon_i \eta_i$ so γ is a circuit in $\Gamma(\mathcal{Y},Y)$. But $\Gamma(\mathcal{Y},Y)$ has no circuits, hence $\gamma = 0$ and there must be some edge followed by its negative. Upon examining this edge we find that $e_i = e_{i+1}$ in Y and $h_i \in \mathcal{U}_{\varepsilon_i}(e_i)\, \mathcal{U}(e_i)\, \mathcal{U}_{\varepsilon_i}(e_i)^{-1}$. Hence if $i \neq n$, then either (σ,τ) or (σ,τ') is not reduced. So $i = n$ and $\varepsilon_n e_n = \lambda_m f_m$. An obvious induction shows that $\sigma = \sigma'$. □

We want to give some examples in which the above yields particularly satisfying results. Along the way we will derive some classical results concerning amalgamated products.

As a first case consider the example of a <u>bouquet of loops</u>, which is a graph having a single vertex. For example, the following is a bouquet of loops in which $|E(Y)| = 4$.

$Y =$ 

Let $(\mathcal{Y},Y)$ be the trivial graph of groups with Y a bouquet of loops. There is a unique maximal subtree of Y consisting of $V(Y)$ and no edges, and an essentially unique reduced realization in $\pi(\mathcal{Y},Y)$. The presentation given in (6.2) says that $\pi(\mathcal{Y},Y)$ is generated by the set $\{\mathcal{U}_{+1}(e) \mid e \in E(Y)\}$ and no relations. Thus $\pi(\mathcal{Y},Y)$ is a free group on $|E(Y)|$ generators. (Note that $|E(Y)|$ need not be finite.)

The result of (7.7) says that each element g of $\pi(\mathcal{Y},Y;\ v)$ is uniquely expressible as

$$g = (\sigma,\tau) = Q_{e_n}^{\varepsilon_n} Q_{e_{n-1}}^{\varepsilon_{n-1}} \cdots Q_{e_1}^{\varepsilon_1}$$

with (σ,τ) reduced. In this case (σ,τ) is reduced if and only if, $e_i = e_{i+1}$ implies $\varepsilon_i = \varepsilon_{i+1}$. Evidently τ and τ' are σ-equivalent if and only if $\tau = \tau'$. Now, for any cardinal n a free group on n generators can be expressed as the fundamental group of a connected, faithful graph of groups - just use a bouquet of loops with n edges. Using the isomorphism of (7.3) we have the following well-known result.

(7.8) Let G be a free group on generators $\{x_i\}_{i\in I}$, then each element g of G is uniquely expressible as

$$g = x_{i_1}^{n_1} x_{i_2}^{n_2} \dots x_{i_r}^{n_r}$$

<u>where</u> $0 \neq n_j \in \mathbb{Z}$, $i_j \neq i_{j+1}$. □

As a second example consider the graph of groups $(\mathcal{Y},Y)$ with

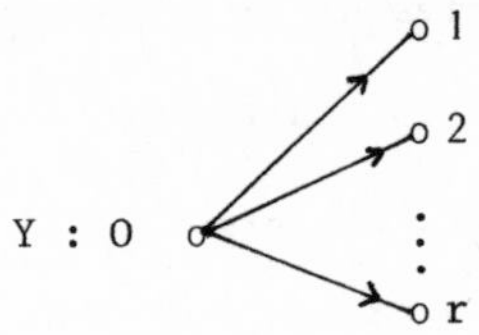

We denote the vertices by the integers $0,\dots,r$ and the edges by e_i with $e_i^{+1} = i$. Let A be a group and suppose that the following hold:

(i) $\mathcal{Y}(o) = A$,

(ii) $\mathcal{Y}(e_i) = A$, $1 \leq i \leq r$,

(iii) $\mathcal{Y}_{-1}(e_i) = id$,

(iv) $\mathcal{Y}_{+1}(e_i)$ is monic.

Under these assumptions there is a unique maximal subtree of Y and an essentially unique reduced realization $\mathcal{R}$ of $(\mathcal{Y},Y)$ in $\pi(\mathcal{Y},Y)$. The presentation given by (6.2) says that $\pi(\mathcal{Y},Y)$ is generated by the vertex groups $\mathcal{R}(i)$, $0 \leq i \leq r$ together with relations that identify the image of A in the different vertex-groups. The fundamental group $\pi(\mathcal{Y},Y)$ is called the <u>amalgamated product of the</u> $\mathcal{R}(i)$ <u>over</u> $\mathcal{R}(o)$ (also the free product with amalgamation).

From (7.7) each element g of $\pi(\mathcal{Y},Y;\,0)$ can be written as

$$g = g_n\, Q_{e_n}^{\varepsilon_n}\, g_{n-1}\, Q_{e_{n-1}}^{\varepsilon_{n-1}} \cdots g_o = (\sigma,\tau)$$

with $g_n, g_o \in \mathcal{U}(0)$ and (σ,τ) reduced. From the structure of Y and the fact that (σ,τ) is reduced we see that the following hold:

(i) $n = 2m$ is even,

(ii) $\sigma = (0, e_{i_1}, i_1, -e_{i_1}, 0, e_{i_2}, i_2, -e_{i_2}, \ldots, e_{i_m}, i_m, -e_{i_m}, 0)$,

(iii) $i_j \neq i_{j+1}$, $1 \leq j \leq m-1$.

Now choose coset representatives $K_i = \{g_i^j \mid 1 \leq j \leq t_i\}$, $1 \leq i \leq r$, for $\mathcal{U}_{+1}(e_i)\, \mathcal{U}(e_i)\, \mathcal{U}_{+1}(e_i)^{-1}$ in $\mathcal{U}(i)$. We claim that τ is σ-equivalent to $\tau' = (g_o', \ldots, g_n')$ satisfying

(*) (i) $g_{2i}' = 1$, $1 \leq i \leq m$,

(ii) $g_{2i-1}' \in K_i$, $1 \leq i \leq m$,

and that τ' is uniquely determined by these properties. This can easily be done by induction on m.

First, if $n = 0 = m$, this is immediate. Suppose that $m > 0$. By induction $\tilde{g} = (\tilde{\sigma},\tilde{\tau})$, with $\tilde{\sigma} = (0, e_{i_2}, i_2, -e_{i_2}, \ldots, e_{i_m}, i_m, -e_{i_m}, 0)$ and $\tilde{\tau} = (g_2, \ldots, g_n)$, may be assumed to satisfy (i) and (ii) of (*). Now consider $g_2\, Q_{e_{i_1}}^{-1}\, g_1 Q_{e_{i_1}}\, g_o$. Since $\mathcal{Y}_{-1}(e_i) = \mathrm{id}$ we have that $\mathcal{U}_{-1}(e_i) = 1$ for each i, so

$$g_2 = 1 \cdot \mathcal{U}_{-1}(e_{i_1}) g_2 \, \mathcal{U}_{-1}(e_{i_1})^{-1} .$$

Choose $g_1' \in K_{i_1}$ and $a_1 \in A$ subject to

$$\mathcal{U}_{+1}(e_{i_1}) g_2 \, \mathcal{U}_{+1}(e_{i_1})^{-1} g_1 = g_1' \, \mathcal{U}_{+1}(e_{i_1}) a_1 \, \mathcal{U}_{+1}(e_{i_1})^{-1} .$$

Both g_1' and a_1 are uniquely determined by this condition. It is now easy to compute that

$$g_2 \, Q_{e_{i_1}}^{-1} \, g_1 \, Q_{e_{i_1}} \, g_0 = 1 \cdot Q_{e_{i_1}}^{-1} \, g_1' \, Q_{e_{i_1}} \, g_0'$$

with $g_0' = \mathcal{U}_{-1}(e_{i_1}) a_1 \mathcal{U}_{-1}(e_{i_1})^{-1} g_0$, and that (g_0, g_1, g_2) and $(g_0', g_1', 1)$ are σ'-equivalent with $\sigma' = (0, e_{i_1}, i_1, -e_{i_1}, 0)$ and the sequence of elements (a_1, g_2). Thus the claim holds. We mention that the above arguments do not depend on the fact that n is finite, in fact n can be any cardinal number. The reader is invited to work out the minor details. Further, if $(\mathcal{Y}, Y)$ is the trivial graph of groups, then the resulting fundamental group is just the free group on r generators considered above.

As in the case of the free group above we can use the isomorphism of (7.3) to get the following result:

(7.9) <u>Let</u> G <u>be the amalgamated product of the groups</u> $\{G_i\}_{i \in I}$ <u>over the subgroup</u> A. <u>Let</u> $K_i = \{g_{ij} \mid j \in J_i\}$ <u>be coset representatives for</u> A <u>in</u> G_i. <u>Then each element</u> g <u>of</u> G <u>is uniquely expressible as</u>

$$g = a\, g_{i_1 j_1}\, g_{i_2 j_2} \cdots g_{i_n j_n}$$

where $a \in A$, $g_{i_k j_k} \in K_{i_k}$, $j_k \in J_{i_k}$, and $i_k \neq i_{k+1}$. □

The one final case we wish to consider is that of a graph of groups $(\mathcal{Y},Y)$ with $Y = v$  e . Again there is a unique maximal subtree and an essentially unique reduced realization $\mathcal{R}$ of $(\mathcal{Y},Y)$ in $\pi(\mathcal{Y},Y)$. The presentation given in (6.2) says that $\pi(\mathcal{Y},Y)$ is generated by the vertex-group $\mathcal{Y}(v)$ together with one further generator having the effect of conjugating the one image of $\mathcal{Y}(e)$ in $\mathcal{Y}(v)$ onto the other. If $(\mathcal{Y},Y)$ is the trivial graph of groups, then $\pi(\mathcal{Y},Y)$ is clearly just the free group on one generator which we considered in (7.8). The general case is known as the <u>HNN-construction</u>. We leave it to the reader to determine a normal form for the elements of $\pi(\mathcal{Y},Y)$ (see 9. below).

8. Exercises and further results.

In this section we want to give some further properties of groups acting on graphs. We do this by way of a number of exercises, the more difficult ones accompanied by hints as to their solution.

The first exercise shows that the assumption that $(\mathcal{Y},Y)$ be faithful is crucial if $(\mathcal{Y},Y)$ is to have reasonable properties.

1. Let $Y = v \circ\!\xrightarrow{\;e\;}\!\circ\, w$ with $\mathcal{Y}(v) \simeq A_5$, $\mathcal{Y}(w) \simeq \mathbb{Z}_2$, $\mathcal{Y}(e) \simeq \mathbb{Z}_2 \times \mathbb{Z}_2$, and $\mathcal{Y}_{+1}(e)$ an epimorphism, $\mathcal{Y}_{-1}(e)$ an isomorphism onto a Sylow 2-subgroup of $\mathcal{Y}(v)$. Then $\pi(\mathcal{Y},Y) = 1$.

Let Γ be a tree. G a group acting on Γ. We first consider subgroups of G and their fixed points.

2. Let H be a finite subgroup of G, then H fixes a vertex. In particular, if $\Gamma = \Gamma(\mathcal{Y},Y)$ for some faithful graph of groups, and $G = \pi(\mathcal{Y},Y)$, then H lies in some conjugate of some vertex-group.

3. Suppose G is generated by elements $\{a_i\}_{i\in I}$, $\{b_j\}_{j\in J}$ and put A and B to be the subgroups of G generated by the a_i and b_j, respectively. Suppose that A and B have fixed-points on Γ, and that each element a_ib_j has a fixed-point. Then G has a fixed-point.

4. Following Serre, we say that G has the property (FA) if G has fixed-points on <u>any</u> tree on which it acts. This property is closely related to <u>not</u> being the fundamental group of a graph of groups $(\mathcal{Y},Y)$ with

$Y = \circ\!\!\longrightarrow\!\!\circ$. In particular, show the following.

Suppose that G is countable. Then G has the property (FA) if and only if the following three conditions are satisfied.

(i) G is not the fundamental group of a faithful graph of groups $(\mathcal{Y},Y)$ with $Y = \circ\!\!\longrightarrow\!\!\circ$,

(ii) G has no quotient isomorphic to $\mathbb{Z}$,

(iii) G is finitely generated.

(Hint: The implications (FA) $\Rightarrow$ (i) and (ii) are easy. (FA) $\rightarrow$ (iii): As G is countable, it is the union of an increasing sequence of finitely generated subgroups, $G_1 \subset G_2 \subset G_3 \dots \subset G_n \subset \dots$. Form a graph Γ with vertex-set $V(\Gamma) = \bigcup_i G_i \diagdown G$ and edge-set $E(\Gamma) = \{(G_i g, G_{i+1} g)\}$ with obvious incidence maps. Γ is a tree, so property (FA) yields a fixed point. This shows that $G = G_i$ for some i.

Assume G satisfies (i), (ii), (iii), and acts on the tree Γ. It follows from (ii) and (6.4) that Γ/G is a tree. Let $(\mathcal{Y},Y)$ be the graph of groups determined by the operation of G on Γ as in (3.6). Thus $G = \pi(\mathcal{Y},\Gamma/G)$ and since G is finitely generated there is a subtree T' of Γ/G such that $\pi(\mathcal{Y}|_{T'},T') = G$. Choose T' minimal with this property. If T' has no edges, then G has a fixed point. Otherwise show that G violates condition (i).)

We assume now that $(\mathcal{Y},Y)$ is a connected faithful graph of groups, $G = \pi(\mathcal{Y},Y)$. The next few exercises consider the other extreme of having fixed-points, namely that of being free.

5. Any subgroup of G which is disjoint from each conjugate of each vertex-group is free.

6. (Schreier) Suppose G is free and H is a subgroup of G. Then H is free. If, moreover, rank G and $|G:H|$ are finite, then

$$\text{rank } H = |G:H|((\text{rank } G)-1) + 1.$$

7. (Serre) Suppose Y is finite and that the vertex-groups are finite. Then for any free subgroup H of finite index in G:

$$\frac{\text{rank } H - 1}{|G:H|} = \sum_{E(Y)} \frac{1}{|\mathcal{Y}(e)|} - \sum_{V(Y)} \frac{1}{|\mathcal{Y}(v)|}$$

The following result is remarkably useful.

8. Let $\mathcal{R}$ be the realization of $(\mathcal{Y},Y)$ in G via (6.2). For each $v \in V(Y)$ let H_v be a subgroup of $\mathcal{R}(v)$ and suppose that $\langle H_v, \mathcal{R}_{+1}(e) \mid v \in V(Y),\ e \in E(Y)\rangle = G$ (note that $\mathcal{R}_{-1}(e) = 1$). Assume that for each edge e of Y

$$\mathcal{R}_{+1}(e)(H_{e^{-1}} \cap \mathcal{R}(e))\mathcal{R}_{+1}(e)^{-1} = H_{e^{+1}} \cap \mathcal{R}_{+1}(e)\mathcal{R}(e)\mathcal{R}_{+1}(e)^{-1} .$$

Then $H_v = \mathcal{R}(v)$.

(Hint: For $e \in E(Y)$ set $H_e = H_{e^{-1}} \cap \mathcal{R}(e)$. Put $X' = \bigcup_Y H_y \setminus G$ and define a graph structure on X'. There is an obvious surjective graph morphism $X' \longrightarrow \Gamma(\mathcal{Y},Y)$ which is locally injective by our hypothesis, that is, $\varphi|_{\Delta(\alpha)}$ is injective for each $\alpha \in V(X')$. Hence, it is an isomorphism by (3.4) and the result holds.)

Normal forms now come into play.

9. Let $Y =$  so that G is an HNN-construction. Find a normal form for elements of $\pi(\mathcal{Y},Y)$ similar to that of (7.9).

(Hint: Replace Y by $Y' =$  where x is a new vertex with vertex-group the same as the previous edge-group. Define the new edge-groups so that $\pi(\mathcal{Y}',Y') \simeq \pi(\mathcal{Y},Y)$.)

10. Let $n \in \mathbb{N}$ and T_n be the following tree:

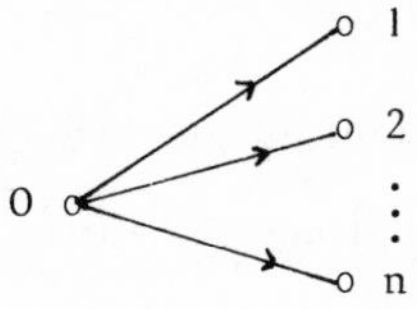

Suppose that G is the amalgamated product of the vertex-groups G_i over the edge-group G_0 (see (7.9)). For I a subset of $\{1,\ldots,n\}$ let $G_I = \langle G_i \mid i \in I \rangle$. (If $I = \phi$, set $G_I = G_0$.) Then for any two subsets I,J of $\{1,\ldots,n\}$

$$G_I \cap G_J = G_{I \cap J}.$$

A homological result is next.

11. Suppose $(\mathcal{Y},Y)$ is a faithful graph of groups with $Y = \circ\!\!\longrightarrow\!\!\circ$. Let N be a free normal subgroup of $\pi = \pi(\mathcal{Y},Y)$, $\bar{\Gamma} = \Gamma(\mathcal{Y},Y)/N$. Let $C_1 = \mathbb{Z}[E(\bar{\Gamma})]$, $C_0 = \mathbb{Z}[V(\bar{\Gamma})]$ and define $\delta : C_1 \longrightarrow C_0$, $\varepsilon : C_0 \longrightarrow \mathbb{Z}$ by

$$\delta(\bar{e}) = \bar{e}^{+1} - \bar{e}^{-1} ,$$

$$\varepsilon(\bar{v}) = 1 ,$$

and extend δ,ε to homomorphisms. Lastly set $H_1 = \text{Kern}\ \delta$.

(a) $0 \longrightarrow H_1 \longrightarrow C_1 \overset{\delta}{\longrightarrow} C_0 \overset{\varepsilon}{\longrightarrow} \mathbb{Z} \longrightarrow 0$ is exact.

(b) H_1 and N/N' are isomorphic π/N-modules.

The last result is an interesting application of the above to fusion in groups. Let $Y = v\ \circ\!\!\overset{e}{\longrightarrow}\!\!\circ\ w$ and $(\mathcal{Y},Y)$ a faithful graph of groups. Let $\mathcal{R}$ be a connected realization of $(\mathcal{Y},Y)$ in a group G.

12.(a) Let Q be a subgroup of $\mathcal{R}(e)$. Set $\mathcal{K} = \{Q^g \mid g \in G\}$, $\mathcal{K}(y) = \mathcal{K} \cap \mathcal{R}(y)$, $y \in Y$. Each set $\mathcal{K}(y)$ is the union of $\mathcal{R}(y)$-conjugacy classes $K_y^1 \cup \ldots \cup K_y^{n_y} = \mathcal{K}(y)$. Define a graph $\Phi = \Phi(G,Q)$ with vertices $V(\Phi) = \{K_v^i, K_w^j\}$ and edges $\{K_e^h\}$. The incidence maps are determined by inclusion in the obvious way and orientation is chosen so that $(K_e^h)^{+1} \in \{K_w^i\}$. Lastly, let $\Gamma = \Gamma(\mathcal{R})$ and put Γ^Q to be the set of fixed-points of Q on Γ. Then

$$\Gamma^Q/N_G(Q) \simeq \Phi .$$

(b) Generalize (a) to a faithful graph of groups with Y an arbitrary graph.

References

1. W. Dicks, Groups, trees, and projective modules, Springer Lecture Notes, no. 790.

2. J. Rotman, Introduction to homological algebra, Academic Press, New York, 1979.

3. J.P. Serre, Trees, Springer Verlag, New York, 1980.

Part II

Weak (B,N)-pairs of rank 2

A.L. Delgado and B. Stellmacher

1. Introduction and notation.

In the first part of this book it was described how to obtain generators and relations for a group G acting on a connected graph Γ in terms of the isotropy subgroups of G and the fundamental group of Γ. In this second part we show how, under some circumstances, the structure of the isotropy subgroups may be determined from the quotient graph Γ/G and the local operation of the isotropy subgroups on the graph Γ. This method was introduced in [7], a fundamental paper for all investigations in this area.

The bulk of this part of the book is an application of this method to the problem of classifying all groups with a weak (B,N)-pair of rank 2. The definition of a weak (B,N)-pair as well as the statement of our result (Theorem A) will be given in chapter 4.

The reader interested in general techniques for the analysis of groups via a collection of subgroups should read Chapters 2 and 3. Chapter 4 is a discussion of Theorem A including examples of groups satisfying the hypotheses. Chapters 5 - 13 contain the proof of Theorem A. Of independent interest may be Chapter 5 which describes some of the internal structure of rank 1 Chevalley groups and has a number of results related to their GF(p)-modules, and also Chapter 6 which includes some pushing-up results. In Chapter 3 and later in Chapter 14 a geometric construction of (B,N)-pairs of rank 2 is described. Some problems, exercises, and generalizations are discussed in Chapter 16.

If in a graph Γ each pair of distinct vertices is joined by at most one edge and Γ contains no loops, then it is convenient to identify

Γ with its set of vertices $V(\Gamma)$. In this case adjacency in Γ is viewed as a symmetric, non-reflexive relation.

Γ will now denote a graph in the above sense. We define some general concepts adapting the notation to the present situation. For $\delta \in \Gamma$, that is δ a vertex of Γ, let $\Delta(\delta)$ be the set of vertices in Γ, adjacent to δ. Note that $\delta \notin \Delta(\delta)$. A path γ of length n in Γ is an (ordered) $(n+1)$-tuple $(\alpha_o,\ldots,\alpha_n)$ of vertices of Γ so that $\alpha_i \in \Delta(\alpha_{i+1})$ for $0 \leq i \leq n-1$, and $\alpha_i \neq \alpha_j$ for $i \neq j$ and $(i,j) \neq (0,n)$. This definition is a little more restrictive than that given in the first part of the book. With $d(\ ,\ )$ we denote the usual distance metric on Γ.

Let G be a group acting on Γ, that is G operates on the vertices and respects adjacency, and let $\delta \in \Gamma$. Then G_δ is the stabilizer of δ in G, and $G_\delta^{(n)}$ (for $n \geq 1$) is the normal subgroup of G_δ fixing all vertices of distance at most n from δ. We use the exponential notation for the action of G on Γ, i.e. δ^g is the image of δ under g. We say that δ^g is G-conjugate to δ.

2. Amalgams of rank n.

(2.0) Hypothesis. In this chapter let G be a group, $I = \{1,\ldots,n\}$ and P_i, $i \in I$, proper subgroups of G. Suppose G satisfies:

(A1) $G = \langle P_1,\ldots,P_n \rangle \neq \langle P_i \mid i \in J \rangle$ for $J \subseteq I$ and $J \neq I$,

(A2) $B = P_i \cap P_j$, $i \neq j$, is independent of $i,j \in I$,

(A3) no non-trivial normal subgroup of G is contained in B.

We say that G is an amalgam of rank n of the subgroups $P_1,\ldots,P_n$ over B, or simply an amalgam of rank n.

Let Γ be the (right) coset graph of G with respect to $P_1,\ldots,P_n$ and B, i.e.

$$\Gamma = \{P_i x, Bx \mid i \in I, x \in G\},$$

where two vertices α and β of Γ are adjacent, if and only if they are distinct and $\alpha \subseteq \beta$ or $\beta \subseteq \alpha$. Then G operates on Γ by right multiplication.

There is a natural relation between Γ and the graphs investigated in Part I of this book. In the language and notation developed there consider the graphs Y_n, for $n \in \mathbb{N}$.

Y_n :
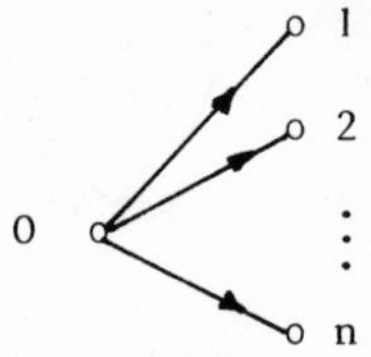

and let $(\mathcal{Y},Y_n)$ be the graph of groups defined by

$$\begin{aligned}
\mathcal{Y}(0) &= B, \\
\mathcal{Y}(i) &= P_i \\
\mathcal{Y}((0,i)) &= B \qquad 1 \leq i \leq n. \\
\mathcal{Y}_{\pm 1}((0,i)) &= Id
\end{aligned}$$

Let $\mathcal{R}$ be the realization of $(\mathcal{Y},Y_n)$ in G and $\Gamma(\mathcal{R})$ be the graph of the realization. From the structure of Y_n we see that $\Gamma(\mathcal{R})$ has no loops and that each pair of distinct vertices is joined by at most one edge. Hence we may identify $\Gamma(\mathcal{R})$ with its set of vertices and incidence becomes a symmetric, non-reflexive relation. With this identification, it is easy to see that $\Gamma(\mathcal{R})$ and the graph Γ defined above are identical.

From chapter 3 on we will investigate amalgams of rank 2. In this case a vertex of type Bx is adjacent to precisely two other vertices, namely P_1x and P_2x. To simplify notation further in this case we will identify Γ with the vertices $\{P_1x, P_2x \mid x \in G\}$ where two such vertices are adjacent if and only if they are distinct and have non-empty intersection. A moments thought will reveal that this identification is compatible with the G-action on Γ.

We assume now that G satisfies (2.0) and Γ is the associated graph. The following is immediate from the definition of Γ.

(2.1) Let $\delta_i, \beta \in \Gamma$ with $\delta_i = P_i x$ for $i \in I$ and $\beta = By$. Then the following hold:

(a) $G_{\delta_i} = P_i^x$ and $G_\beta = B^y$.

(b) The edge-stabilizers in G are conjugate to B.

(c) $\Delta(\delta_i) \simeq P_i/B$ as a G-set; in particular G_{δ_i} is transitive on $\Delta(\delta_i)$.

(d) $G_\beta = G_\beta^{(1)}$ and $G_{\delta_i}^{(1)} = \bigcap_{g \in G_{\delta_i}} B^{xg}$.

(e) $(\Delta(\beta) \cup \{\beta\})^G = \Gamma$.

(f) $\langle G_\delta \mid \delta \in \Delta(\beta) \rangle = G$. □

(2.2) G operates faithfully on Γ.

Proof: Let $g \in G$ fix Γ pointwise. Then $g \in B$ and $\langle g^G \rangle \leq B$. Now (A3) implies that $g = 1$. □

(2.3) Γ is connected.

Proof: Let $\beta \in \Gamma$ so that $\beta = B$ and let Γ_o be the connected component of Γ containing β. Then (2.1)(a) implies $\langle P_1, \ldots, P_n, B \rangle \leq N_G(\Gamma_o)$ and (A1) together with (2.1)(e) yields $\Gamma = \Gamma_o$. □

(2.4) Let $\beta \in \Gamma$ with $\beta = Bx$ and L_δ be a subgroup of G_δ for $\delta \in \Delta(\beta)$ which operates transitively on $\Delta(\delta)$. Then for $L = \langle L_\delta \mid \delta \in \Delta(\beta) \rangle$ the following hold:

(a) $\Gamma/G \simeq \Gamma/L$.

(b) $G = LB^x$.

(c) If $L_\delta \cap L_{\delta'} = M$ is independent of the choice of $\delta,\delta' \in \Delta(\beta)$, $\delta \neq \delta'$, then L is an amalgam of rank n of the subgroups $L_1,\ldots,L_n$ over M.

Proof: By the Frattini argument and (2.1), we have that $G_\delta = L_\delta B^x$. Now (a) and (b) follow from (2.1)(d), (e) and (f). Assertion (c) is an immediate consequence of (2.2). □

One obvious question which the above result raises is - with L_δ, L as above - whether $L_\delta = L_\delta$ (?!). This is not as silly as it looks. L_δ on one side is the subgroup of G_δ given by hypothesis, while that on the other is the stabilizer in L of the vertex δ. It is not a priori clear that the two L_δ's should be equal. There is, however, one important case where this is in fact true, namely, when Γ is a tree. Hence, at least in this instance the apparent ambiguity of the notation L_δ is removed. Although we will not use this fact, we will record it here.

(2.5) Assume Γ is a tree. Let $\beta \in \Gamma$ with $\beta = Bx$ and L_δ be a subgroup of G_δ, for $\delta \in \Delta(\beta)$. Assume that $M = L_\delta \cap L_{\delta'}$ is independent of the choice of $\delta,\delta' \in \Delta(\beta)$, $\delta \neq \delta'$, and set $L = \langle L_\delta \mid \delta \in \Delta(\beta)\rangle$. Then the stabilizer of the vertex δ in L is just L_δ.

Proof. This is (Part I, (8.8)). □

The most generally useful and easy to apply result is the following harmless looking lemma.

(2.6) Let $\beta \in \Gamma$ with $\beta = Bx$, and let N be a subgroup of B^x such that $N_{G_\delta}(N)$ operates transitively on $\Delta(\delta)$ for $\delta \in \Delta(\beta)$. Then $N = 1$.

Proof: Set $L = \langle N_{G_\delta}(N) \mid \delta \in \Delta(\beta) \rangle$. From (2.4)(a), we have $\Gamma/L \simeq \Gamma/G$. Hence N lies in the kernel of the operation of G on Γ, and (2.2) yields $N = 1$. □

(2.7) Let $\beta \in \Gamma$ with $\beta = B$. For each $\delta \in \Delta(\beta)$ let N_δ be a normal subgroup of G_δ which operates transitively on $\Delta(\delta)$. Set $N = \langle N_\delta \mid \delta \in \Delta(\beta) \rangle$ and $B_o = \prod_{\delta \in \Delta(\beta)} (B \cap N_\delta)$. Then N is normal in G, and N is an amalgam of the subgroups $N_\delta B_o$, $\delta \in \Delta(\beta)$, over B_o.

Proof: It follows from (2.4)(b) that N is normal in G. In view of (2.4)(c) it remains to show that $B_o = N_\delta B_o \cap N_{\delta'} B_o$ for $\delta, \delta' \in \Delta(\beta)$, $\delta \neq \delta'$, but this is easy to check. □

The comments after (2.4) apply equally well to (2.7).

There are very few results at this level of generality which yield structural information about $P_1, \ldots, P_n$. The only exceptions we know are based on techniques developed by Wielandt and Bender.

Definition: Let G be a finite group. A component E of G is a perfect subnormal subgroup E of G so that $E/Z(E)$ is simple. Let $E(G)$ be the group generated by all components of G (i.e. $E(G) = 1$, if there are no components in G) and $F^*(G) = F(G)E(G)$ where $F(G)$ is the Fitting subgroup of G.

We assume the reader to be familiar with the properties of $E(G)$ and $F^*(G)$ (for reference see [2]). In particular, one should know that any subnormal p-group is contained in $F(G)$ and $C_G(F^*(G)) \leq F^*(G)$.

Apart from properties of $F^*(G)$ only the $P \times Q$ - Lemma [8, 5.3.4] is used in the next lemmata. We begin with the following hypothesis.

(2.8) Hypothesis. For the lemmata (2.9) and (2.10) let $P_1,\ldots,P_n$ be finite groups and N be a non-trivial subgroup of B.

In the following proofs we use the graph theoretical language developed in this chapter. Let $\beta \in \Gamma$ with $\beta = B$. Note that $\Delta(\beta) = \{\delta_i \mid \delta_i = P_i,\ i \in I\}$. We define $N_\beta = N$ and choose a fixed conjugate N_μ of N_β in G_μ for every $\mu \in \Gamma$ with $d(\mu,\beta) = 2$.

(2.9) *Suppose that N is subnormal in P_i, for some $i \in I$, and $C_{P_i}(E(N)) \leq B$. Then $F^*(P_i) = F^*(B)$.*

Proof: We may assume that $N_\beta = E(N_\beta)$ and $P_i = G_\delta$ for some $\delta \in \Delta(\beta)$. Note that $N_\beta \leq F^*(G_\delta)$. Then $F^*(G_\delta) = N_\beta C_{F^*(G_\delta)}(N_\beta)$ and the hypothesis implies $F^*(G_\delta) \leq F^*(B)$; in particular $E(G_\delta) \leq E(B)$ and $F(G_\delta) \leq F(B)$.

Since $N_\mu \leq E(G_\delta)$ for $\mu \in \Delta(\delta)$, we get
$F(B) \leq \bigcap_{\mu\in\Delta(\delta)} (G_\delta \cap G_\mu) = G_\delta^{(1)}$ and thus $F(B) \leq F(G_\delta^{(1)}) \leq F(G_\delta)$. We conclude that $F(B) = F(G_\delta)$.

Now let K be a component of $E(B)$. If $K \not\leq F^*(G_\delta)$, then K centralizes $F^*(G_\delta)$ which contradicts $C_{G_\delta}(F^*(G_\delta)) \leq F^*(G_\delta)$. Hence, we have $E(B) = E(G_\delta)$ and therefore $F^*(B) = F^*(G_\delta)$. □

(2.10) Let p be a prime, and suppose that N is subnormal in P_i for some $i \in I$, and $G_{P_i}(O_p(N)) \leq B$. Then $O^p(F^*(P_i)) = O^p(F^*(B))$.

Proof: We may assume that $N_\beta = O_p(N_\beta)$ and $P_i = G_\delta$ for some $\delta \in \Delta(\beta)$. As in (2.9) we get $O^p(F^*(G_\delta)) \leq O^p(F^*(B))$. Since $N_\beta O^p(F^*(B))$ operates on $O_p(G_\delta)$, the $P \times Q$-Lemma together with the hypothesis yield $[O_p(G_\delta), O^p(F^*(B))] = 1$. Hence $O^p(F^*(B))$ centralizes N_μ for $\mu \in \Delta(\delta)$, and we get $O^p(F^*(B)) \leq G_\delta^{(1)}$. It follows that
$O^p(F^*(B)) \leq O^p(F^*(G_\delta^{(1)})) \leq O^p(F^*(G_\delta))$, and the assertion holds. □

(2.11) Let P_i, $i \in I$, be finite, $I_o \subseteq I$ and $|I_o| = 2$. Suppose that for $k \in I_o$ there exists a non-trivial subgroup N_k in B so that for each $m \in I$ and any prime q:

(i) N_k is subnormal in P_m,

(ii) $C_{P_m}(E(N_k)) \leq P_k$, if $E(N_k) \neq 1$, and $C_{P_m}(O_q(N_k)) \leq P_k$, if $O_q(N_k) \neq 1$.

(iii) $N_{P_i}(O_q(P_j)) \leq P_j$ for $i,j \in I_o$, if $O_q(P_j) \neq 1$.

Then there exists a prime p so that $F^*(P_i)$, $i \in I$, and $F^*(B)$ are p-groups.

Proof: In view of (A1) and (A3) it suffices to show that there exists a fixed prime p so that

$$(*) \qquad O^p(F^*(P_m)) = O^p(F^*(B)) \quad \text{for every} \quad m \in I .$$

Note that without loss we may assume that either $N_i = O_{p_i}(N_i)$ for some prime p_i, or $N_i = E(N_i)$. We apply (2.9) and (2.10). If $E(N_i) = N_i$ for each $i \in I_o$, then $F^*(B) = F^*(P_m)$ for all $m \in I$, whence $F^*(B) = 1$ by (2.6) which is impossible. If $E(N_i) = N_i$, $O_q(N_j) = N_j$, $\{i,j\} = I_o$, then $F^*(P_m) = F^*(B)$ for $m \neq i$ by (2.9), and $O^p(F^*(P_i)) = O^p(F^*(B))$ by (2.10). But then the desired result (*) follows. Hence, we may assume that $N_i = O_p(N_i)$ and $N_j = O_q(N_j)$, and furthermore, again by (2.10), that $p \neq q$. Now (2.10) applied to N_i and P_m for $m \in I \setminus \{i\}$ shows (*) for $m \neq i$. Hence, it suffices to prove

$$(**) \qquad O^p(F^*(P_i)) = O^p(F^*(B)) .$$

As usual we translate this configuration into graph theoretic language. There exist different vertices $\delta, \lambda \in \Delta(\beta)$ with $P_i = G_\delta$ and $P_j = G_\lambda$. We define $N_\delta = N_i$ and $N_\lambda = N_j$ and choose $N_\mu \in N_\lambda^G \cap G_\mu$ for $\mu \in \lambda^G$ and $d(\mu,\delta) = 2$.

Since $N_\lambda \leq O_q(G_\delta)$ we have by (ii) $O_p(G_\delta) \leq B \leq G_\lambda$. Set $P = N_{O_p(G_\lambda)}(O_p(G_\delta))$. Then (iii) implies $P \leq B$ and thus $P \leq O_p(B)$. On the other hand, again by (ii) $C_{O_q(G_\delta)}(N_\lambda) \leq O_q(B)$ and we get $[P, C_{O_q(G_\delta)}(N_\lambda)] = 1$. Now the $P \times Q$-Lemma applied to $P \times N_\lambda$ and $O_q(G_\delta)$ yields $[P, O_q(G_\delta)] = 1$. In particular, $[P, N_\mu] = 1$ where $d(\mu,\delta) = 2$ and

$\mu \in \lambda^G$, and (ii) implies that

$$P \leq \bigcap_{x \in G_\delta} B^x = G_\delta^{(1)} .$$

Since $G_\delta^{(1)} \leq G_\lambda$, it follows that $P \leq O_p(G_\delta^{(1)})$, whence $P \leq O_p(G_\delta)$. Now the definition of P yields $O_p(G_\lambda) \leq O_p(G_\delta)$.

Define $N = O_p(G_\lambda)$, i.e. $N = O_p(P_j)$. Then N, P_i fulfil the hypotheses of (2.10) whence (**) holds. □

We now apply (2.11) to derive some information about vertex-stabilizers.

(2.12) Let $\beta \in \Gamma$ with $\beta = Bx$. Suppose that G_δ is finite for $\delta \in \Delta(\beta)$ and $N_{G_\lambda}(M) \leq G_\delta$ for every non-trivial normal subgroup M of G_δ and $\delta, \lambda \in \Delta(\beta)$.

Then $G_\mu^{(3)} \neq 1$ for at most one vertex $\mu \in \Delta(\beta)$, or there exists a prime p so that $F^*(G_\delta)$, $\delta \in \Delta(\beta)$, and $F^*(G_\beta)$ are p-groups.

Proof: Without loss we may assume (after conjugation) that $\beta = B$. Then $G_\beta = B$ and $\{G_\delta \mid \delta \in \Delta(\beta)\} = \{P_i \mid i \in I\}$. In addition, we may assume that $G_\delta^{(3)} \neq 1 \neq G_\lambda^{(3)}$ for two different vertices $\delta, \lambda \in \Delta(\beta)$.

Let $G_\delta = P_i$ and $G_\lambda = P_j$, and define $N_i = G_\delta^{(3)}$ and $N_j = G_\lambda^{(3)}$. Then N_i and N_j and $I_o = \{i,j\}$ fulfil the hypotheses of (2.11), and the assertion follows. □

Lemma (2.11) shows that for $P_1,\ldots,P_n$ finite groups, in many cases there exists a "characteristic" prime p so that

$$C_{P_i}(O_p(P_i)) \leq O_p(P_i), \quad i \in I .$$

In the next chapter we will make this part of the hypothesis.

We close this chapter with two final definitions which will come into play later in chapter 4.

Definition: (1) Suppose G satisfies (2.0), and let X be the amalgamated product of the groups P_i over B. Then a group H is locally isomorphic to G (with respect to the groups P_i) if there exists a free normal subgroup Y of X with $X/Y \simeq H$.

(2) Suppose G satisfies (2.0) with respect to $P_1,\ldots,P_n$ and H satisfies (2.0) with respect to $Q_1,\ldots,Q_n$. Then G and H are parabolic isomorphic if (after a possible permutation of indices) $P_i \simeq Q_i$, $P_i \cap P_j \simeq Q_i \cap Q_j$, $i,j \in I$.

Clearly, being locally isomorphic implies being parabolic isomorphic. It is easy to construct examples in which the converse is not true.

3. Amalgams of rank 2.

(3.0) Hypothesis. In this chapter we assume Hypothesis (2.0) and $|I| = 2$. In addition, there exists a fixed prime p so that

(B1) P_1 and P_2 are finite subgroups and $Syl_p(B) \subseteq Syl_p(P_i)$, $i = 1,2$.

(B2) $C_{P_i}(O_p(P_i)) \leq O_p(P_i)$, $i = 1,2$.

We use the notation introduced in chapter 2. But - as it was mentioned there - we use a "streamlined" version of the graph Γ. For clarity we repeat its definition.

Γ is the set of (right) cosets of G with respect to P_1 and P_2, and two vertices of Γ are adjacent, if and only if they are distinct and have non-empty intersection.

A moments thought will show that the coset graph defined with respect to P_1, P_2, B as in chapter 2 may be indentified with the graph above in a way which is compatible with the G-action. When we speak of an edge of Γ we will mean a pair of adjacent vertices. Note that, by definition of Γ, the stabilizer of an edge is the intersection of the two defining vertex-stabiliers. It is now clear that the results of chapter 2 also hold for this new version of Γ. For the sake of completeness we list the basic results proven there.

We note that the Lemmata (3.1) - (3.3) do not rely on (B1) and (B2), and Lemma (3.4) does not rely on (B2).

(3.1) Let δ_1 and δ_2 be adjacent vertices of Γ.

(a) Γ is connected.

(b) G operates faithfully on Γ.

(c) Vertex-stabilizers of Γ are G-conjugate to P_1 or P_2.

(d) Edge-stabilizers of Γ are G-conjugate to B.

(e) There is a $g \in G$ such that $\{G_{\delta_1^g}, G_{\delta_2^g}\} = \{P_1, P_2\}$.

(f) For $\delta_i = P_i$, $\Delta(\delta_i)$ is isomorphic to P_i/B as a G_{δ_i}-set, and G_{δ_i} operates transitively on $\Delta(\delta_i)$. □

(3.2) Let δ_1 and δ_2 be adjacent vertices of Γ, and suppose for $i = 1,2$ that L_{δ_i} is a subgroup of G_{δ_i} operating transitively on $\Delta(\delta_i)$. Set $L = \langle L_{\delta_1}, L_{\delta_2} \rangle$.

(a) L operates edge-transitively on Γ.

(b) $G = (G_{\delta_1} \cap G_{\delta_2})L$.

(c) If N is a normal subgroup of each of L_{δ_1} and L_{δ_2}, then $N = 1$.

(d) If L_{δ_i} is normal in G_{δ_i}, $i = 1,2$, and $B_0 = (B \cap L_{\delta_1})(B \cap L_{\delta_2})$, then L is an amalgam of $L_{\delta_1}B_0$ and $L_{\delta_2}B_0$ over B_0.

(e) If Γ is a tree, then the stabilizer of δ_i in L is just $L_{\delta_i}B_0$. □

(3.3) Let Δ be a connected graph and H a group of automorphisms of Δ which operates edge- but not vertex-transitively on Δ, and let α and β be adjacent vertices in Δ. Then H is an amalgam of H_α and H_β over $H_\alpha \cap H_\beta$, and the corresponding coset graph $\Gamma(H, H_\alpha, H_\beta)$ is H-isomorphic to Δ.

Proof: The map $H_\delta g \longrightarrow \delta^g$, $g \in H$ and $\delta = \alpha, \beta$, is a bijection which is compatible with the operation of H. Since (H_α, H_β) is mapped onto an edge, this map is a H-isomorphism. Set $M = \langle H_\alpha, H_\beta \rangle$. The subgraph Δ_o with vertex set $\alpha^M \cup \beta^M$ is a connected component of Δ and so $\Delta_o = \Delta$. Now a Frattini argument shows $H = M(H_\alpha \cap H_\beta) = M$, and the result follows. □

Lemma (3.3) together with (3.1) shows that Hypothesis (3.0) is equivalent to the following.

Hypothesis. Let Γ be a connected graph, p a fixed prime and G a group of automorphisms of Γ so that for $\delta \in \Gamma$:

(a) G operates edge- but not vertex-transitively on Γ,

(b) G_δ is finite,

(c) $C_{G_\delta}(O_p(G_\delta)) \leq O_p(G_\delta)$,

(d) $Syl_p(G_\alpha \cap G_\beta) \subseteq Syl_p(G_\alpha) \cap Syl_p(G_\beta)$ for α, β adjacent in Γ.

From now on it is up to the reader's taste to think of Γ as an arbitrary connected graph or the more concrete coset graph.

We now define some notions related to the operation of G on Γ.

Let $\gamma = (\alpha_o, \ldots, \alpha_n)$ be a path in Γ. We say that γ is regular, if G_γ operates transitively on $\Delta(\alpha_o) \setminus \{\alpha_1\}$ and on $\Delta(\alpha_n) \setminus \{\alpha_{n-1}\}$. In addition, we define

$$s = \min\{|\gamma| \mid \gamma \text{ is non-regular}\},$$

$$r = \max\{|\gamma| \mid \gamma \text{ is regular}\}.$$

Here $|\gamma|$ denotes the length of γ. It is sometimes necessary to stress the group and graph with respect to which the parameters r and s are defined, in which case we write $s(\Gamma,G)$ and $r(\Gamma,G)$.

Let α and β be adjacent vertices in Γ and let K be a subgroup of $G_\alpha \cap G_\beta$. We say that K is a <u>Cartan subgroup</u> of G, if

(*) there exists $t_\delta \in N_{G_\delta}(K) \setminus (G_\alpha \cap G_\beta)$ so that $t_\delta^2 \in K$, $\delta = \alpha,\beta$.

In this case we set $N = \langle t_\alpha, t_\beta \rangle K$ and $W = N/K$, and say that W is a <u>Weyl group</u> with respect to K.

If K is a Cartan subgroup in $G_\alpha \cap G_\beta$ with Weyl group W, the induced subgraph T with vertex set $T = \{\alpha^w, \beta^w \mid w \in W\}$ is called an <u>apartment</u> with respect to K and W. If we want to stress the correspondence between K, W and T we write $W(K)$, $T(K)$ or $T(K,W)$.

Note that a Cartan subgroup is not assumed to be non-trivial; if there are involutions in $G_\delta \setminus G_\delta^{(1)}$, $\delta = \alpha,\beta$, then $K = 1$ is a Cartan subgroup. Note further that there is nothing unique about Cartan subgroups.

(3.4) (a) s <u>is finite</u>.

(b) $r \geq s-1$.

(c) <u>For</u> $n \leq s$ G <u>operates transitively on paths of length</u> n <u>whose initial vertices are in the same</u> G-<u>orbit</u>.

<u>Proof</u>: The first claim follows from the finiteness of vertex-stabilizers in G, the others are easy consequences of the definition of r and s. □

If we are only interested in the structure of P_1 and P_2, we may assume without loss that G is the amalgamated product of P_1 and P_2 with respect to B. In this case Γ is a tree (see Part I, (4.4)). The next two lemmata refer to this situation.

(3.5) Suppose that Γ is a tree, and α,β are adjacent vertices in Γ. Let K be a Cartan subgroup in $G_\alpha \cap G_\beta$ with Weyl group $W = W(K)$ and apartment $T = T(K,W)$. Then

(a) T is a subgraph of valence 2 in Γ,

(b) K fixes the vertices in T,

(c) for every $\delta \in T$ there exists $w \in W$ which is a reflection on T and fixes δ.

Proof: Let $N = \langle t_\alpha, t_\beta \rangle K$ as in the definition so that $W = N/K$, and let w_δ be the image of t_δ, $\delta = \alpha,\beta$. Set $\tau = w_\alpha w_\beta$, then every vertex in T is conjugate to α or β by an element of $\langle\tau\rangle$. In addition, for every $n \in \mathbb{N}$

$$\gamma_n = (\beta^{\tau^{-n}}, \alpha^{\tau^{-(n-1)}}, \ldots, \alpha^{\tau^{-1}}, \beta^{\tau^{-1}}, \alpha, \beta, \alpha^{\tau}, \beta^{\tau}, \ldots, \alpha^{\tau^n}, \beta^{\tau^n})$$

is a path. Since there are no closed paths in Γ, (a) follows.

Assertion (b) is obvious.

Note that w_α reflects γ_n and fixes α for every $n \in \mathbb{N}$. Thus, w_α' is a reflection on T. The analogous argument shows that w_β is a reflection fixing β. Now (c) follows after conjugation with $\langle\tau\rangle$. □

Our definition above may suggest to the reader some connection to groups having a (B,N)-pair and to generalized n-gons. We will try to make this connection clear. We first formulate two further properties.

Let T be an apartment in Γ and s the above defined parameter. We say:

T fulfils the <u>uniqueness condition</u>, if every path of length s in Γ is contained in a unique G-conjugate of T, and

T fulfils the <u>exchange condition</u>, if for every path $(\delta_{-(s-1)},\ldots,\delta_o,\ldots,\delta_{s-1})$ of length $2(s-1)$ in T and every $1 \neq x \in G_{(\delta_o,\ldots,\delta_{s-1})}$ there exists $y \in G_{(\delta_{-(s-1)},\ldots,\delta_o)}$ so that $\delta_{-(s-1)}^{xy} = \delta_{s-1}$.

Note that by (3.4) every path of length s is contained in at least one conjugate of T. We will refer to any conjugate of T as an apartment.

Let T and T^g be different apartments so that $|T \cap T^g| = s-1$. Then the uniqueness condition and the definition of s say that T and T^g are conjugate under the stabilizer of $T \cap T^g$, whereas the uniqueness condition together with the exchange condition say that there is a further apartment T^h, $h \in G$, so that

$$|T \cap T^h| = |T^g \cap T^h| = s-1$$

and

$$T \cap T^h \cap T^g \text{ consists of a single point.}$$

A finite bipartite graph is a <u>generalized</u> n-<u>gon</u>, $n \geq 2$, if the following hold:

(a) Any two vertices are joined by a path of length at most n.

(b) There is at most one path of length less than n between any pair of vertices.

(c) Every vertex lies on a closed path of length $2n$.

(3.6) *Let* Γ *be a tree and* K *a Cartan subgroup of* G *with apartment* $T = T(K)$. *Suppose that* T *fulfils the uniqueness and exchange conditions and that* $s \geq 3$. *Then there exists an equivalence relation* $\approx$ *on* Γ *which is compatible with the operation of* G *so that*

(1) $\tilde{\Gamma} = \Gamma/\approx$ *is a generalized* $(s-1)$-gon,

(2) $G_{\tilde{\Gamma}} \cap G_\delta = 1$, $\delta \in \Gamma$,

(3) $G/G_{\tilde{\Gamma}}$ *is finite*.

Proof: For $\delta_1, \delta_2 \in \Gamma$, $\delta_1 \neq \delta_2$, let $\gamma(\delta_1,\delta_2)$ denote the unique path in Γ from δ_1 to δ_2. We write $\delta_1 \sim \delta_2$, if and only if $d(\delta_1,\delta_2) = 2(s-1)$ and there exists a G-conjugate of T containing $\gamma(\delta_1,\delta_2)$. Then $\sim$ extends to an equivalence relation $\approx$ on Γ. Obviously, this equivalence relation is compatible with the operation of G. We write $\tilde{\delta}$ for the equivalence class containing $\delta \in \Gamma$ and $\tilde{\Gamma} = \Gamma/\approx$. Then two vertices $\tilde{\alpha}, \tilde{\beta} \in \tilde{\Gamma}$ are adjacent, if $\tilde{\alpha} \neq \tilde{\beta}$ and there exists $\alpha \in \tilde{\alpha}$ and $\beta \in \tilde{\beta}$ so that $\alpha \in \Delta(\beta)$.

Let $\Delta = \{\delta \mid d(\delta,\alpha) \leq s-1\}$ for fixed $\alpha \in \Gamma$. We claim that Δ contains at least one representative for each equivalence class of Γ. This will then show that $\tilde{\Gamma}$ is finite and condition (a) in the definition of a generalized $(s-1)$-gon holds.

Let σ be any vertex of $\Gamma \setminus \Delta$ and $\gamma = \gamma(\sigma,\alpha)$. The definition of s and (3.4) imply that there exists $g \in G$ so that $\sigma \in T^g$ and

$|T^g \cap \gamma| \geq s$. Thus, there exists $\sigma' \in T^g$ so that $\sigma \sim \sigma'$ and $d(\sigma,\alpha) > d(\sigma',\alpha)$. An obvious induction now shows the above claim.

Next we show that no two different vertices at distance less than $2(s-1)$ are equivalent. Suppose that $\delta,\delta' \in \Gamma$ with $0 \neq d(\delta,\delta') < 2(s-1)$ and $\delta' \in \tilde{\delta}$. Then there exist vertices $\delta_o,\ldots,\delta_k \in \Gamma$ with $\delta_o = \delta$, $\delta_k = \delta'$ and $\delta_i \sim \delta_{i+1}$, $0 \leq i \leq k-1$. Let k be minimal with these properties. Choose $m \in \{1,\ldots,k-1\}$ with $d(\delta_o,\delta_m)$ maximal and $d(\delta_o,\delta_{m+1}) < d(\delta_o,\delta_m)$ and set $\lambda = \gamma(\delta_{m-1},\delta_m) \cap \gamma(\delta_{m+1},\delta_m) = (\alpha,\ldots,\delta_m)$. By the choice of m we have $|\lambda| = s-1$. Note that by the uniqueness condition there exists $x \in G_\lambda$ so that $\delta_{m+1} = \delta^x_{m-1}$. Hence applying the exchange condition to the path $\gamma(\delta_{m-1},\delta_m)$ and x there exists $y \in G_{(\delta_{m-1},\ldots,\alpha)}$ so that $\delta_m^{y^{-1}} = \delta_{m+1}$. It follows that $\delta_{m-1} \sim \delta_{m+1}$ which contradicts the minimality of k. This implies that condition (b) of a generalized $(s-1)$-gon is satisfied. The uniqueness condition yields condition (c). Having verified the three axioms, (1) holds, and as we have seen above (3) holds as well.

Let $x \in G_{\tilde{\Gamma}} \cap G_\delta$, and $\lambda \in \Delta(\delta)$. Since $G_\Gamma = 1$ we may assume that $\lambda^x \neq \lambda$. But this implies $d(\lambda,\lambda^x) = 2 = 2(s-1)$ and $s = 2$, a contradiction. □

A group D is said to have a (B,N)-<u>pair</u>, if there exist subgroups B,N of D satisfying:

(a) $D = \langle B,N \rangle$ and $B \cap N$ is normal in N,

(b) $W = N/B \cap N$ is generated by a set of involutions S,

(c) $BsB \cdot BwB \subseteq BwB \cup BswB$, for $s \in S$ and $w \in W$,

(d) $sBs \neq B$, for $s \in S$.

The cardinality of S is the <u>rank</u> of the (B,N)-pair.

(3.7) Assume the hypothesis and notation of (3.6). Then $G/G_{\tilde{\Gamma}}$ has a (B,N)-pair of rank 2.

Proof: Let α,β be adjacent vertices in Γ. According to (3.1) we may assume that $B = G_\alpha \cap G_\beta$ and $K \leq B$. We choose the following notations:

$N = \langle t_\alpha, t_\beta \rangle K$, t_δ as in the above definition,

$W = N/K$,

$S = \{w_\alpha, w_\beta\}$, w_δ the image of t_δ, $\delta = \alpha,\beta$,

$T = T(K,W)$,

$\tilde{G} = G/G_{\tilde{\Gamma}}$.

We use the $\sim$-convention for the images of elements and subsets of G in $\tilde{G}$. We want to show that $\tilde{B}$ and $\tilde{N}$ is a (B,N)-pair of $\tilde{G}$ with respect to the set $\tilde{S}$.

Note that $s \geq 3$ and (3.4) imply that G_δ is doubly transitive on $\Delta(\delta), \delta = \alpha,\beta$. Hence [8, 2.7.3] yields

$$G_\delta = B \cup Bt_\delta B, \ \delta = \alpha,\beta \ ;$$

in particular, $G = \langle B,N \rangle$ and $B \neq t_\delta^{-1} B t_\delta$. In addition, the operation of N on $\tilde{T}$ shows that $\tilde{B} \cap \tilde{N} = \tilde{K}$.

Since S is a set of involutions, it remains to prove condition (c) in the definition of a (B,N)-pair.

Pick $\gamma = (\lambda_o,\ldots,\lambda_{s-1}) \subseteq T$ so that $\{\lambda_o,\lambda_1\} = \{\alpha,\beta\}$ and set $R = G_\gamma$. Since γ is regular by (3.4), we have that R operates transitively on $\Delta(\lambda_o) \setminus \{\lambda_1\}$. It follows that $B = B_oR$ where $B_o = G_{\lambda_{-1}} \cap B$ and $\lambda_{-1} \in (T \cap \Delta(\lambda_o)) \setminus \{\lambda_1\}$. Note that t_{λ_o} normalizes B_o since $B_o = G_{\lambda_{-1}} \cap G_{\lambda_o} \cap G_{\lambda_1}$.

Let $n \in N$ and $t = t_{\lambda_o}$. Suppose first that $R^n \leq B$. Then

$$BtB \cdot BnB = BtB_oRnB = BtnB .$$

Next suppose that $R^n \not\leq B$. We now use the factor graph $\tilde{\Gamma}$. By (3.6) $\tilde{T}$ is a closed path of length $2(s-1)$ on which $\tilde{N}$ operates, and $\tilde{R}$ fixes one half of this path whereas $\tilde{R}^{\tilde{t}}$ fixes the other. It follows that $\tilde{R}^{\tilde{t}\tilde{n}} \leq \tilde{B}$ and as above

$$\tilde{B}\tilde{t}\tilde{B} \cdot \tilde{B}\tilde{t}\tilde{n}\tilde{B} = \tilde{B}\tilde{n}\tilde{B} .$$

This implies that

$$\tilde{B}\tilde{t}\tilde{B} \cdot \tilde{B}\tilde{n}\tilde{B} = \tilde{B}\tilde{t}\tilde{B} \cdot \tilde{B}\tilde{t}\tilde{B} \cdot \tilde{B}\tilde{t}\tilde{n}\tilde{B} \subseteq (\tilde{B} \cup \tilde{B}\tilde{t}\tilde{B})\tilde{B}\tilde{t}\tilde{n}\tilde{B} \subseteq \tilde{B}\tilde{t}\tilde{n}\tilde{B} \cup \tilde{B}\tilde{n}\tilde{B} ,$$

and $\tilde{B}$ and $\tilde{N}$ is a (B,N)-pair of $\tilde{G}$. □

Suppose that the hypotheses and notation of (3.6) again hold. Assume further that $G_\alpha \cap G_\beta = B$ is p-closed where p is the prime of (3.0). By (3.7) we know that $G/G_{\tilde{\Gamma}}$ has a rank 2 (B,N)-pair. The groups satisfying these conditions have been classified by Fong and Seitz [6]. They show that $G/G_{\tilde{\Gamma}}$ is isomorphic to one of

$$L_3(p^n), PSp_4(p^n), U_4(p^n), U_5(p^n), G_2(p^n), {}^3D_4(p^n), {}^2F_4(2^n) .$$

(Actually, one first needs to reduce to a certain minimal configuration, but we will not stop to make this precise.) We see that the uniqueness and exchange conditions impose very strong restrictions on the group G. In the next chapter we will consider hypotheses (Hypothesis A) on G which will often lead to the conclusion that these two conditions in fact hold. The exceptional cases lead to fascinating configurations associated to finite sporadic groups.

We will now introduce the group theoretic machinery which will be at the heart of the analysis of the groups P_1 and P_2. We set the following:

Notation. Let $\delta \in \Gamma$ and define

$$\begin{aligned}
\Delta^{(i)}(\delta) &= \{\lambda \in \Gamma \mid d(\delta,\lambda) \leq i\}, \\
Q_\delta &= O_p(G_\delta), \\
Z_\delta &= \langle \Omega_1(Z(T)) \mid T \in Syl_p(G_\delta)\rangle, \\
V_\delta^{(i)} &= \langle Z_\lambda \mid \lambda \in \Delta^{(i)}(\delta)\rangle, \quad i \geq 1, \\
V_\delta &= V_\delta^{(1)} .
\end{aligned}$$

We furthermore define the following basic parameters.

$$\begin{aligned}
b_\delta &= \min_{\delta' \in \Gamma} \{d(\delta,\delta') \mid Z_\delta \nleq G_{\delta'}^{(1)}\}, \\
b &= \min_{\delta' \in \Gamma} \{b_{\delta'}\}.
\end{aligned}$$

Notice that as G operates edge-transitively, $b = \min\{b_\alpha, b_\beta\}$ for any pair of adjacent vertices α,β. Thus we are allowed to choose α and β so that

$$b_\alpha = b \leq b_\beta$$

and

$$G_\alpha \cap G_\beta = B \quad \text{and} \quad \{G_\alpha, G_\beta\} = \{P_1, P_2\}.$$

Choose $\alpha' \in \Gamma$ so that

$$d(\alpha,\alpha') = b \quad \text{and} \quad Z_\alpha \not\leq G_{\alpha'}^{(1)} ,$$

and let γ be a path of length b from α to α'. We label the vertices of γ by

$$\gamma = (\alpha,\alpha+1,\ldots,\alpha+b) = (\alpha'-b,\ldots,\alpha'-1,\alpha') ,$$

i.e. $\alpha+i$ (resp. $\alpha'-i$) is the unique vertex in γ with $d(\alpha,\alpha+i) = i$ (resp. $d(\alpha',\alpha'-i) = i$). Furthermore, from (3.1)(f) we may assume that $\beta = \alpha+1$, if $b \geq 1$.

This notation will remain fixed throughout the remainder of this book.

(3.8) <u>For</u> $\delta \in \Gamma$

(a) $Z_\delta \leq Z(Q_\delta) \cap V_\delta$,

(b) $Q_\delta \leq G_\delta^{(1)}$; <u>in particular</u>, $b \geq 1$,

(c) $Z_{\alpha'} \leq G_\alpha$ <u>and</u> $[Z_\alpha, Z_{\alpha'}] \leq Z_\alpha \cap Z_{\alpha'}$,

(d) $Z_\alpha \neq \Omega_1(Z(T))$, $T \in Syl_p(G_\alpha)$,

(e) <u>if</u> $S \in Syl_p(B)$ <u>and</u> $\Omega_1(Z(S))$ <u>is centralized by a subgroup</u> R <u>of</u> G_β <u>which operates transitively on</u> $\Delta(\beta)$, <u>then</u> $Z(G_\alpha) = 1$.

<u>Proof</u>: It is clear by (B1) and (B2) that (a) and (b) hold, and (c) follows from the minimality of the parameter b.

Assume that $Z_\alpha = \Omega_1(Z(T))$, $T \in Syl_p(G_\alpha)$. Since Z_α is normal in G_α, we get $Z_\alpha \leq Z_\beta$ and so $Z_\beta \nleq G_{\alpha'}^{(1)}$, contradicting the minimality of b. Hence (d) holds.

It remains to prove (e). By (B2) $\Omega_1(Z(G_\alpha)) \leq \Omega_1(Z(S))$ and hence $\Omega_1(Z(G_\alpha))$ is centralized by $\langle G_\alpha, R\rangle$. Now (3.2)(c) yields $\Omega_1(Z(G_\alpha)) = 1 = Z(G_\alpha)$. □

The above defined parameter b is closely related to the structure of P_1 and P_2. Roughly speaking, b describes the number of non-central chief factors in $O_p(G_\alpha)$. We give an easy example.

(3.9) <u>Example</u>. <u>Suppose</u> $p = 2$ <u>and</u> $P_i/O_2(P_i) \simeq \Sigma_3$, $i = 1,2$. <u>Assume that</u> $b = 1$. <u>Then</u> $P_i \simeq \Sigma_4$ <u>or</u> $\Sigma_4 \times C_2$ <u>for</u> $i = 1,2$.

<u>Proof</u>: We have that $G_\delta/Q_\delta \simeq \Sigma_3$, $Q_\delta = G_\delta^{(1)}$, $\delta \in \Gamma$, and $B \in Syl_2(G_\alpha) \cap Syl_2(G_\beta)$.

Note that $(\alpha, \alpha+1) = (\alpha'-1, \alpha')$ and $\beta = \alpha'$ by the choice of β; in particular, it follows that $Z_\alpha \nleq Q_\beta$, $Z_\alpha Q_\beta = B$ and $|B/Q_\alpha| = 2$.

Since Z_α is normal in B and central in Q_α, it follows that

$$[Z_\alpha, Q_\beta, Q_\beta] = 1 \quad \text{and} \quad [Z_\alpha, Q_\beta] \leq \Omega_1(Z(B)) \leq Z_\beta.$$

Set $L = \langle Z_\alpha^{G_\beta} \rangle$. Then L operates transitively on $\Delta(\beta)$ and $[Q_\beta, L] \leq Z_\beta$. If $Z_\beta \leq Q_\alpha$, then $[Q_\beta, L, L] = 1$, and $O^2(L)$ centralizes Q_β contradicting (B2). Hence $Z_\beta \nleq Q_\alpha$, and we have $Z_\beta(Q_\alpha \cap Q_\beta) = Q_\beta$ and $Z_\alpha(Q_\alpha \cap Q_\beta) = Q_\alpha$. Thus $\phi(Q_\alpha) = \phi(Q_\alpha \cap Q_\beta) = \phi(Q_\beta)$, and (3.2)(c) implies $\phi(Q_\alpha) = \phi(Q_\beta) = 1$. Now for $\delta = \alpha, \beta$ the operation of G_δ on Q_δ and the fact that $Q_\alpha \cap Q_\beta \leq \Omega_1(Z(B))$ yields $Q_\delta = Z_\delta$.

We now have $|Z_\alpha / Z_\alpha \cap Z_\beta| = |Z_\beta / Z_\alpha \cap Z_\beta| = 2$ and so $|Z_\delta / \Omega_1(Z(G_\delta))| = 4$, $\delta = \alpha, \beta$. On the other hand, again (3.2)(c) implies $\Omega_1(Z(G_\alpha)) \cap \Omega_1(Z(G_\beta)) = 1$. We conclude that $|\Omega_1(Z(G_\delta))| \leq 2$ and either $G_\delta \simeq \Sigma_4$ or $C_2 \times \Sigma_4$, $\delta = \alpha, \beta$. □

Note the exceptionally strong role played by (3.2)(c) in the above argument. It is essentially the only tool we will have to force contradictions and control the structure of the groups G_α and G_β.

Note further that the minimal parabolic subgroups of $L_3(2)$ and $Sp_4(2)$ containing of fixed Borel subgroup provide examples for the two cases in the conclusion of (3.9).

We now demonstrate how to determine the parameter b. As an example we have chosen the situation Goldschmidt considered in [7], so this example is highly non-trivial. In fact, apart from the identification of P_1 and P_2, it is a good pattern for the proof of Theorem A (stated in chapter 4).

(3.10) Example. Suppose that $p = 2$ and $P_i/O_2(P_i) \simeq \Sigma_3$, $i = 1,2$. Then one of the following holds:

(a) $[Z_\alpha, Z_{\alpha'}] \neq 1$ and $b = 1$ or 2.

(b) $[Z_\alpha, Z_{\alpha'}] = 1$ and $b = 3$.

(c) There exist $\delta, \lambda \in \{\alpha, \beta\}$ so that for $L = \langle Q_\delta^{G_\lambda} \rangle$:

(c1) $Q_\delta \in Syl_2(L)$,

(c2) $L/O_2(L) \simeq \Sigma_3$,

(c3) no non-trivial characteristic subgroup of Q_δ is normal in L.

Proof: Let (G_α, G_β) be a counter example. As in (3.9) we have $B \in Syl_2(G_\alpha) \cap Syl_2(G_\beta)$, $B = Q_\alpha Q_\beta$ and $Q_\delta = G_\delta^{(1)}$ for $\delta \in \Gamma$.

(1) Let $\{\delta, \lambda\} = \{\alpha, \beta\}$ and $x \in G_\lambda \setminus G_\delta$, then $\langle x, Q_\delta \rangle = G_\lambda$.

Set $L = \langle x, Q_\delta \rangle$, and note that $|B/Q_\delta| = 2$ and $LB = G_\lambda$. If $L \neq G_\lambda$, then L fulfils (c1) and (c2). Let C be a characteristic subgroup of Q_δ which is normal in L. Since L is transitive on $\Delta(\lambda)$ and C is normal in G_δ, it follows from (3.2)(c) that $C = 1$, and (G_α, G_β) is not a counter example.

(2) Let $\Omega_1(Z(B)) < U \leq Z_\alpha$, then $C_{G_\alpha}(U) = Q_\alpha$ and $N_{G_\beta}(U) = B$.

The first claim follows from (3.8)(d) and the action of G_α on Z_α. In addition, we have $[B, Z_\alpha] \leq \Omega_1(Z(B))$ and thus $B \leq N_{G_\beta}(U)$. If $B \neq N_{G_\beta}(U)$, then U is normal in G_β and $Q_\alpha \leq C_{G_\beta}(U) \nleq B$. Now (1) implies $U \leq \Omega_1(Z(G_\beta)) \leq \Omega_1(Z(B))$, a contradiction.

(3) $Z_\alpha = \Omega_1(Z(B))\Omega_1(Z(G_\alpha \cap G_{\alpha-1}))$, $\alpha-1 \in \Delta(\alpha) \setminus \{\beta\}$.

This is again an immediate consequence of the operation of G_α on Z_α.

It should be noted that all these statements may also be applied to conjugate vertices; for example, (1) holds for any pair of adjacent vertices.

Let s be the parameter defined earlier in this chapter, and let $\hat{\gamma} = (\delta_0,\ldots,\delta_s)$ be a path of length s.

(4) Suppose that $s < b$. If $G_{\hat{\gamma}} \not\leq Q_{\delta_s}$, then δ_s is conjugate to α'.

There exists a non-regular path $\tilde{\gamma} = (\tilde{\delta}_0,\ldots,\tilde{\delta}_s)$ of length s so that $G_{\tilde{\gamma}}$ is not transitive on $\Delta(\tilde{\delta}_s) \setminus \{\tilde{\delta}_{s-1}\}$. Since $|\Delta(\tilde{\delta}_s)| = 3$ we get $G_{\tilde{\gamma}} \leq G_{\tilde{\delta}_s}^{(1)} = Q_{\tilde{\delta}_s}$. Now (3.4)(c) implies that $\tilde{\delta}_s$ is not conjugate to δ_s. Hence, since there are two G-orbits of vertices in Γ, it suffices to show that $\tilde{\delta}_s$ is conjugate to $\alpha'-1$.

Again by (3.4)(c) we may assume that $\tilde{\gamma} \subseteq \gamma$ and $\tilde{\delta}_s \in \{\alpha'-1,\alpha'\}$. On the other hand, $Z_\alpha \leq G_\gamma \leq G_{\tilde{\gamma}}$ and $Z_\alpha \not\leq Q_{\alpha'}$, and $\tilde{\delta}_s = \alpha'-1$ follows.

We now divide the proof into two separate cases.

I. <u>The case</u> $[Z_\alpha, Z_{\alpha'}] \neq 1$.

Set $R = [Z_\alpha, Z_{\alpha'}]$. We have $|Z_\alpha/Z_\alpha \cap Q_{\alpha'}| = |Z_{\alpha'}/Z_{\alpha'} \cap Q_\alpha| = 2$, $Z_\alpha Q_{\alpha'} = G_{\alpha'} \cap G_{\alpha'-1}$ and $Z_{\alpha'} Q_\alpha = B$. The operation of G_δ on Z_δ yields:

(5) $|Z_\delta/\Omega_1(Z(G_\delta))| = 4$ and $\Omega_1(Z(B)) = R\Omega_1(Z(G_\alpha))$ for $\delta = \alpha, \alpha'$.

Since (G_α, G_β) is a counter example, we have $b > 2$. In particular V_α is elementary abelian. Pick $\alpha-1 \in \Delta(\alpha) \setminus \{\beta\}$ so that

(6) $\langle G_{\alpha-1} \cap G_\alpha, Z_{\alpha'} \rangle = G_\alpha$.

We first show:

(7) $b \leq s$.

By way of contradiction we may assume that $s < b$. Let $\hat{\gamma} = (\alpha'-(s+1), \ldots, \alpha'-1)$. It follows from (4) that $G_{\hat{\gamma}} \leq Q_{\alpha'-1} \leq G_{\alpha'}$.

Note that $Z_{\alpha-1} \leq G_{\hat{\gamma}}$. Thus, we have $Z_{\alpha-1} \leq Z_\alpha Q_{\alpha'}$ and $[Z_{\alpha-1}, Z_{\alpha'}] = R \leq Z_\alpha$ by (3.8)(c), and $Z_\alpha Z_{\alpha-1}$ is normal in G_α by (6).

Suppose first that $Z_{\alpha-1} \not\leq Z_\alpha$. Then $Z_{\alpha-1} \neq \Omega_1(Z(G_{\alpha-1} \cap G_\alpha))$ and as in (2) $C_{G_{\alpha-1}}(Z_{\alpha-1}) = Q_{\alpha-1}$. In particular, $C_{Q_\alpha}(Z_\alpha Z_{\alpha-1}) = Q_\alpha \cap Q_{\alpha-1}$, and $Q_\alpha \cap Q_{\alpha-1}$ is normal in G_α. Set $L = \langle Q_{\alpha-1}^{G_\alpha} \rangle$. Then $L/Q_\alpha \cap Q_{\alpha-1}$ is a central extension of Σ_3, and we conclude that $Q_{\alpha-1} \in Syl_2(L)$, a contradiction to (1).

Suppose now that $Z_{\alpha-1} \leq Z_\alpha$ and thus $Z_\beta \leq Z_\alpha$. Since $B = Q_\alpha Q_\beta$, we get $Z_\beta = \Omega_1(Z(G_\beta))$, and since $Z_{\alpha'} \neq \Omega_1(Z(G_{\alpha'}))$, the vertices β and α' are not conjugate and b is even, i.e. $b \geq 4$. In addition, by (3.8)(e) $Z(G_\alpha) = 1$, and (5) implies

$$Z_\alpha = Z_{\alpha-1} \times Z_\beta \quad \text{and} \quad R = Z_\beta.$$

Since the situation is symmetric in α and α', we get with the same argument $R = Z_{\alpha'-1}$ and

$$R = Z_{\alpha'-1} = Z_\beta.$$

As $b \geq 4$, we have $V_\alpha^{(2)} \leq Q_\alpha$ and $[V_\alpha^{(2)},R] = 1$. On the other hand, the minimality of b implies $V_\alpha^{(2)} \leq G_{\alpha'-2}$; and since $\alpha'-2$ is conjugate to α, we get $Z_{\alpha'-2} = Z_{\alpha'-3} \times R$. Hence, $V_\alpha^{(2)}$ centralizes $Z_{\alpha'-2}$, and by (2) $V_\alpha^{(2)} \leq Q_{\alpha'-2} \leq G_{\alpha'-1}$. But now $V_\alpha^{(2)} \leq G_{\hat{\gamma}} \leq G_{\alpha'}$, and we get as above

$$[V_\alpha^{(2)},Z_{\alpha'}] \leq Z_\alpha.$$

In particular, $[V_{\alpha-1},Z_{\alpha'}] \leq Z_\alpha \leq V_{\alpha-1}$, and by (6) $V_{\alpha-1}$ is normal in G_α, contradicting (3.2)(c). This final contradiction shows (7).

(8) $Z_{\alpha-1} \not\leq Z(G_{\alpha-1})$.

Assume that $Z_{\alpha-1} \leq Z(G_{\alpha-1})$. As in the proof of (7) we get $R = Z_\beta = Z_{\alpha'-1}$ and $V_\alpha^{(2)} \leq Q_{\alpha'-2}$. On the other hand, by (7) there is $\alpha-2 \in \Delta(\alpha-1) \setminus \{\alpha\}$ so that $(\alpha-2,\alpha-1,\alpha,\ldots,\alpha'-2)$ is conjugate to $(\alpha,\ldots,\alpha')$.

Hence, $Z_{\alpha-2} \not\leq Q_{\alpha'-2}$ which contradicts $V_\alpha^{(2)} \leq Q_{\alpha'-2}$.

We now derive a final contradiction in the case $[Z_\alpha, Z_{\alpha'}] \neq 1$. As in the proof of (7) $Z_{\alpha-1} \leq G_{\alpha'}$ leads to $Z_{\alpha-1} \leq Z_\alpha$ and then to a contradiction to (8).

Assume that $Z_{\alpha-1} \not\leq G_{\alpha'}$. Then by (1) $\langle Z_{\alpha-1}, Q_{\alpha'} \rangle = G_{\alpha'-1}$ and so $R \leq Z(G_{\alpha'-1})$, since $b > 2$. On the other hand, from (7) and (3.4)(c) we get, as above,an $\alpha-2 \in \Delta(\alpha-1) \setminus \{\alpha\}$ so that $Z_{\alpha-2} \not\leq Q_{\alpha'-2}$ and $\langle Z_{\alpha-2}, G_{\alpha'-2} \cap G_{\alpha'-1} \rangle = G_{\alpha'-2}$. Hence, again since $b > 2$, it follows that $R \leq Z(G_{\alpha'-2})$, a contradiction to (3.2)(c).

II. <u>The case</u> $[Z_\alpha, Z_{\alpha'}] = 1$.

Since $Z_\alpha \not\leq Q_{\alpha'}$ we have $Z_{\alpha'} = \Omega_1(Z(G_{\alpha'}))$; in particular, by (2) α' is not conjugate to α. It follows that

(9) $Z_\beta = \Omega_1(Z(G_\beta)) = \Omega_1(Z(B))$, and b is odd,

and together with (3.8)(e) that

(10) $Z(G_\alpha) = 1$ and $Z_\alpha = Z_{\alpha-1} \times Z_\beta$, $\alpha-1 \in \Delta(\alpha) \setminus \{\beta\}$.

Assume first that $b = 1$. Then $\beta = \alpha'$ and, since $[Z_\alpha, B] = Z_\beta$, we get $[Q_\beta, O^2(G_\beta)] = 1$, contradicting (B2). Since (G_α, G_β) is a counter example, we have

(11) $b \geq 5$.

In particular, V_β is elementary abelian and again by (9) and (10)

(12) $[V_\beta,Q_\beta] = Z_\beta$.

Set $R = [V_\beta,V_{\alpha'}]$ and $V = V_\alpha^{(2)}$. The minimality of b implies $V \leq G_{\alpha'-2}$, $V_\beta \leq G_{\alpha'}$, $V_{\alpha'} \leq G_\beta$, and by (2) $R \neq 1$. It follows that

(13) $1 \neq R \leq V_\beta \cap V_{\alpha'}$ and $[R,V] = 1$.

The pair $(V_\beta,V_{\alpha'})$ will now play the role $(Z_\alpha,Z_{\alpha'})$ played in the first part of the proof. Note that there is no longer symmetry in β and α', since we do not know, a priori, that $V_{\alpha'} \not\leq Q_\beta$.

(14) $V \not\leq Q_{\alpha'-2}$.

Assume that $V \leq Q_{\alpha'-2}$. Then there exists $\alpha-1 \in \Delta(\alpha)\setminus\{\beta\}$ so that $d(\alpha-1,\alpha'-2) = d(\beta,\alpha') = b-1$ and $V_{\alpha-1} \leq Q_{\alpha'-2}$. Hence, the paths $(\alpha-1,\alpha,\ldots,\alpha'-1)$ and $(\beta,\ldots,\alpha')$ cannot be conjugate, since $Z_\alpha \leq V_\beta$ and thus $V_\beta \not\leq Q_{\alpha'}$. By (3.4)(c) we have $s < b$. Thus, (4) applied to $\gamma_1 = (\alpha'-(s+1),\ldots,\alpha'-1)$ and $\gamma_2 = (\alpha'-(s-1),\ldots,\alpha',\alpha'+1)$, $\alpha'+1 \in \Delta(\alpha')\setminus\{\alpha'-1\}$, yields $V \leq G_{\gamma_1} \leq Q_{\alpha'-1}$ and $V \cap Q_{\alpha'} \leq G_{\gamma_2} \leq Q_{\alpha'+1}$. It follows that $V \leq Z_\alpha Q_{\alpha'}$ and $[V \cap Q_{\alpha'},V_{\alpha'}] = 1$, and

$$[V,V_{\alpha'}] = [Z_\alpha,V_{\alpha'}] = R.$$

If $V_{\alpha'} \nleq Q_\beta$, then V is normal in $\langle G_\alpha \cap G_\beta, V\rangle = G_\beta$ and G_α, contradicting (3.2)(c). If $V_{\alpha'} \leq Q_\beta$, then $[Z_\alpha, V_{\alpha'}] \leq Z_\alpha$ and $V_{\alpha-1}$ is normal in $\langle G_\alpha \cap G_{\alpha-1}, V_{\alpha'}\rangle = G_\alpha$ and $G_{\alpha-1}$, again contradicting (3.2)(c).

(15) $Z_{\alpha'} \cap R = 1$.

Set $R_0 = Z_{\alpha'} \cap R$. By (13) we have $[V,R_0] = 1$. On the other hand, by (1) and (14) $R_0 \leq Z(\langle V,Q_{\alpha'-1}\rangle) = Z(G_{\alpha'-2})$. It follows by (10) that $R_0 \leq Z_{\alpha'-2} \cap Z_{\alpha'} = 1$.

(16) $|R| = |Z_\beta| = 2$ and $|Z_\alpha| = 4$.

From (12) and (15) we get $[V_\beta \cap Q_{\alpha'}, V_{\alpha'}] = 1$ and $|Z_\alpha/C_{Z_\alpha}(V_{\alpha'})| = 2$. Now (2) implies $C_{Z_\alpha}(V_{\alpha'}) = \Omega_1(Z(B))$, and the operation of G_α on Z_α and (10) imply (16).

(17) $R = Z_{\alpha'-2}$.

Note that by (1) and (14) $\langle V,Q_{\alpha'-1}\rangle = G_{\alpha'-2}$. Hence if $R \leq Z_{\alpha'-1}$, then $R \leq Z_{\alpha'-2}$ and (13) and (16) yield (17).

Assume now that $R \nleq Z_{\alpha'-1}$; i.e. $[R,Q_{\alpha'-1}] \neq 1$. Note that $Q_{\alpha'-1} = Z_\alpha(Q_{\alpha'-1} \cap Q_{\alpha'}) = Z_\alpha^t(Q_{\alpha'-1} \cap Q_{\alpha'-2})$ where $t \in G_{\alpha'-1}$ with $(\alpha'-2)^t = \alpha'$ and $\alpha'^t = \alpha'-2$. It follows that $[R,Z_\alpha^t] = 1$ and with (16)

$$[R,Q_{\alpha'-1}] = [R,Q_{\alpha'-2} \cap Q_{\alpha'-1}] = Z_{\alpha'}.$$

If $Q_{\alpha'-1} \cap Q_{\alpha'-2}$ is normal in $G_{\alpha'-2}$, then $[RV_{\alpha'-2}, Q_{\alpha'-1} \cap Q_{\alpha'-2}] = Z_{\alpha'}$

is normal, too, which contradicts (2).

Thus, we have $Q_{\alpha'-2} = (Q_{\alpha'-3} \cap Q_{\alpha'-2})(Q_{\alpha'-1} \cap Q_{\alpha'-2})$ and $[RV_{\alpha'-2}, Q_{\alpha'-2}] = Z_{\alpha'-3}Z_{\alpha'-1}$. It follows that $V_{\alpha'-2} = Z_{\alpha'-3}Z_{\alpha'-1}$ and after conjugation $V_{\alpha'} = Z_{\alpha'-1}Z_{\alpha'+1}$, $\alpha'+1 \in \Delta(\alpha') \setminus \{\alpha'-1\}$. But now by (16) $|V_{\alpha'}/Z_{\alpha'}| = 4$ and $R \leq Z_{\alpha'-1}$, a contradiction.

(18) $V_{\alpha'} = Z_{\alpha'-1}Z_{\alpha'+1}$, $\alpha'+1 \in \Delta(\alpha') \setminus \{\alpha'-1\}$, and $|V_{\alpha'}| = 2^3$.

By (12) and (17) $[G_{\alpha'} \cap G_{\alpha'-1}, V_{\alpha'}] = Z_{\alpha'-1}$ and thus $[G_{\alpha'}, V_{\alpha'}] = Z_{\alpha'-1}Z_{\alpha'+1} = V_{\alpha'}$. Now (18) follows from (16).

(19) $b \geq 7$.

Assume that $b = 5$. We first show that $s \geq 4$. Let $\tilde{\gamma} = (\tilde{\delta}_o, \dots, \tilde{\delta}_s)$ be as in (4). If $s = 2$, then $Q_{\tilde{\delta}_1} \leq Q_{\tilde{\delta}_2}$, contradicting $Q_{\tilde{\delta}_1}Q_{\tilde{\delta}_2} = G_{\tilde{\delta}_1} \cap G_{\tilde{\delta}_2}$. If $s = 3$, then $Q_{\tilde{\delta}_1} \cap Q_{\tilde{\delta}_2} \leq Q_{\tilde{\delta}_s} \cap Q_{\tilde{\delta}_2}$ and so $Q_{\tilde{\delta}_1} \cap Q_{\tilde{\delta}_2}$ is normal in $G_{\tilde{\delta}_2}$. As in (7) we get $Q_{\tilde{\delta}_1} \in Syl_2(\langle Q_{\tilde{\delta}_1}^{G_{\tilde{\delta}_2}} \rangle)$ and a contradiction to (1).

Hence, $s \geq 4$ and by (3.4)(c) all paths of length 4 with initial vertices in β^G are conjugate. In particular, there exists $\alpha'+2 \in \Delta^{(2)}(\alpha')$ so that $V_{\alpha'+2} \nleq Q_{\alpha+3}$. Pick $t \in V_\beta \setminus Q_{\alpha'}$ and $t' \in V_{\alpha'+2} \setminus Q_{\alpha+3}$. Note that by (18) $V_\beta = \langle t \rangle Z_{\alpha+2}$. Set $z = [t^{t'}, t]$ and $z' = [t'^{t}, t']$.

By (17) we get $z \in Z_{\alpha+3}$ and $z' \in Z_{\alpha'}$. If $z = 1$, then $[V_\beta, V_\beta^{t'}] = 1$ which is impossible, since $(\beta, \dots, \beta^{t'})$ is conjugate to $(\beta, \dots, \alpha')$. With the same argument $z' \neq 1$. Hence, we have $\langle z \rangle = Z_{\alpha+3}$ and $\langle z' \rangle = Z_{\alpha'}$. On the other hand, $\langle t, t' \rangle \simeq D_{16}$ and so $z = z'$. It follows from $b = 5$ that $Z_{\alpha+3} = Z_{\alpha+5}$, contradicting (10) (applied to $\alpha+4$).

We now derive a final contradiction. By (14) there exists $\alpha-1 \in \Delta(\alpha)\setminus\{\beta\}$ and $\alpha-2 \in \Delta(\alpha-1)\setminus\{\alpha\}$ so that $Z_{\alpha-2} \nleq Q_{\alpha'-2}$. Hence, the pair $(\alpha-2,\alpha'-2)$ has the same properties as (α,α'), and we can apply (14) to $(\alpha-2,\alpha'-2)$; i.e. $V^{(2)}_{\alpha-2} \nleq Q_{\alpha'-4}$. On the other hand, since $b \geq 7$, $V^{(2)}_{\alpha-2}$ centralizes V_β and thus by (17) $Z_{\alpha'-2}$. It follows that $V^{(2)}_{\alpha-2} \leq C_{G_{\alpha'-4}}(Z_{\alpha'-3}) \leq Q_{\alpha'-4}$ by (2), a contradiction. □

Note that the maximal 2-local subgroups (parabolic subgroups) of $L_3(2)$, $Sp_4(2)$, $G_2(2)$, $G_2(2)'$, M_{12}, and $Aut(M_{12})$ provide examples for the cases (a) and (b) of (3.10). Case (c) describes a pushing up situation which can be shown to lead to a contradiction (see chapter 6).

As an exercise the reader should try to determine the chief factors of P_1 and P_2 or to give a reasonable bound on the order of B, using the method in (3.9) and starting with the information in (a) and (b), respectively.

4. <u>Weak</u> (B,N)-<u>pairs of rank</u> 2.

In this chapter we want to make a number of preparatory remarks which we hope will motivate the hypotheses below and our main result, Theorem A.

Without further ado, we introduce the hypotheses which will hold for the rest of the book.

<u>Hypothesis</u> A. Let P be a fixed prime and G be an amalgam of the finite subgroups P_1 and P_2 over B. Suppose that for $i = 1,2$ there exists a normal subgroup P_i^* in P_i so that

(i) $O_p(P_i) \leq P_i^*$ and $P_i = P_i^* B$,

(ii) $C_{P_i}(O_p(P_i)) \leq O_p(P_i)$,

(iii) $P_i^* \cap B$ is the normalizer of a Sylow p-subgroup of P_i^* and

$$P_i^*/O_p(P_i) \simeq L_2(p^{n_i}), \ SL_2(p^{n_i}), \ U_3(p^{n_i}), \ SU_3(p^{n_i}), \ Sz(2^{n_i}) \ \text{ or } \ D_{10} \quad \text{(and}$$

$p = 2$), Ree(3^{n_i}) or Ree$(3)'$ (and $p = 3$), $n_i \geq 1$.

Recall that the condition that G be an amalgam implies that no non-trivial normal subgroup of G is contained in B.

Lemma (3.3) shows that Hypothesis A is equivalent to the following

<u>Hypothesis</u> A'. Let G be an edge- but not vertex-transitive group of automorphisms of the connected graph Γ with finite vertex-stabilizers. Let α, β be adjacent vertices of Γ and p a fixed prime. Suppose

that for $\delta = \alpha,\beta$ there exists a normal subgroup G^*_δ in G_δ containing $O_p(G_\delta)$, so that (i) - (iii) of Hypothesis A hold for G^*_δ, $G_\alpha \cap G_\beta$, $G^*_\delta/O_p(G_\delta)$ and n_δ in place of P^*_i, B, $P^*_i/O_p(P_i)$ and n_i.

Let $\overline{G^*_\delta} = G^*_\delta/O_p(G_\delta)$. Then in Hypothesis A' the condition (iii), excluding for the moment the cases $\overline{G^*_\delta} \simeq Ree(3)'$ or D_{10}, implies that $G^{*\Delta(\delta)}_\delta$ is the natural doubly transitive permutation representation. For example, if $\overline{G^*_\delta} \simeq SL_2(p^{n_\delta})$, then $G^{*\Delta(\delta)}_\delta$ is isomorphic to the operation of $PSL_2(p^{n_\delta})$ on the "projective line" with $p^{n_\delta}+1$ points. $\overline{G^*_\delta}$ is in each case a group with a rank 1 (B,N)-pair, and $\overline{G^*_\delta \cap G_\rho}$, for $\rho \in \Delta(\delta)$, is the normalizer of a Sylow p-subgroup of $\overline{G^*_\delta}$, and hence a Borel subgroup of $\overline{G^*_\delta}$. A number of the properties of these groups is listed in (5.1).

The group Ree(3) is the extension of $SL_2(8)$ by the Frobenius autormorphism of GF(8) and Sz(2) is the Frobenius group of order 20 (containing D_{10} as a subgroup of index 2). We allow, for the sake of completeness, the possibility that $G^{*\Delta(\delta)}_\delta \simeq Ree(3)'$ (and $p = 3$) or D_{10} (and $p = 2$). None of these cases causes much difficulty and we will only have to consider them explicitly a couple of times.

It follows from Hypothesis A (iii) that $B \cap P^*_i$, $i = 1,2$, is p-closed, so $B_o = (B \cap P^*_1)(B \cap P^*_2)O_p(B)$ is p-closed, too. Set $G_o = \langle P^*_1B_o, P^*_2B_o\rangle$. Then (3.2) implies that G_o is normal in G and G_o is an amalgam of $P^*_1B_o$ and $P^*_2B_o$ over B_o. Clearly, this amalgam again satisfies Hypothesis A. It will be evident from the proof of Theorem A that all the information we need is already in this "subamalgam". We therefore make the following hypotheses.

Hypothesis B. Hypothesis A holds and, in addition,
$B = (B \cap P_1^*)(B \cap P_2^*)O_p(B)$.

Hypothesis B'. Hypothesis A' holds and, in addition,
$G_\alpha \cap G_\beta = (G_\alpha \cap G_\beta^*)(G_\beta \cap G_\alpha^*)O_p(G_\alpha \cap G_\beta)$.

We will say that a group satisfying Hypothesis A or, equivalently, Hypothesis A' is a group with a weak (B,N)-pair of rank 2 (with respect to P_1, P_2 and B).

A number of partial results in the classification of all pairs G_α, G_β satisfying Hypothesis A' (in particular, the case $p = 2$) have been achieved through the combined works of several authors: Goldschmidt [7], Delgado [4], Fan [5], K.H. Schmidt [15], Stellmacher [16], Stroth [17],[18]. The techniques for these classifications are based on the work of Goldschmidt who in [7] treated the case $|\Delta(\delta)| = 3$, for $\delta = \alpha,\beta$. His methods have undergone successive refinements leading to the unified proof which we give here based on the discussion in chapter 3. Our proof is independent of those of the above authors and is essentially self-contained.

Far from being empty, the class of groups which satisfies Hypothesis A' has a very appealing structure. Let Λ be the following set of groups:

$$\Lambda = \{L_3(p^n), PSp_4(p^n), U_4(p^n), U_5(p^n), G_2(p^n), {}^3D_4(p^n), {}^2F_4(2^n), G_2(2)', {}^2F_4(2)', M_{12}, J_2, F_3 \mid p \text{ prime}, n \in \mathbb{N}\}.$$

The first seven families of groups on the list correspond to the classical and twisted Chevalley groups of rank 2 each having a rank 2 (B,N)-pair. They are all simple with the exception of $PSp_4(2)$, $G_2(2)$, and ${}^2F_4(2)$, each of which has a simple commutator subgroup of index 2. The last three groups on the list are, respectively, the Mathieu group on 12 letters, the second sporadic group of Janko (also called the Hall-Janko-group) and the sporadic group discovered by Thompson. They are all simple.

To each of the groups in Λ we can associate a distinguished prime. For the Chevalley groups this prime is, of course, the characteristic p of the defining field. For $G_2(2)'$, ${}^2F_4(2)'$, M_{12} and J_2 the prime is $p = 2$, and for the Thompson group, F_3, the prime is $p = 3$. This prime p is distinguished by the following property. Let $X \in \Lambda$ and $P \in Syl_p(X)$. Set $Y = N_X(P)$. Then there are precisely two subgroups X_1 and X_2 of X so that X satisfies Hypothesis A' with respect to X_1, X_2 and Y.

Let G be any amalgam with $X_i = P_i$, $i = 1,2$, and $Y = B$ which is locally isomorphic to X. We sometimes call G a realization of X_1 and X_2 over Y. In the next table we list some properties of such realizations, where we use the notation of chapter 3.

X	b	r	s	b_β
$L_3(q)$	1	3	4	1
$PSp_4(2^n)$	1	4	5	1
$PSp_4(q)$, q odd	1	4	5	2
$U_4(q)$	1	4	5	2
$U_5(q)$	1	4	5	2
$G_2(q)$, $q \equiv 0(3)$	2	6	7	2
$G_2(q)$, $q \not\equiv 0(3)$	2	6	7	3
${}^3D_4(q)$	2	6	7	3
${}^2F_4(2^n)$	3	8	9	4
$G_2(2)'$	2	6	5	3
J_2	2	6	5	3
$Aut(J_2)$	2	6	5	3
${}^2F_4(2)'$	3	8	7	4
M_{12}	3	8	5	4
$Aut(M_{12})$	3	8	7	4
F_3	5	-	7	6

The parameters b,r,s,b_β will turn out to be independent of the realization in all cases <u>except</u> for an F_3-amalgam. Here r cannot be specified a priori. We comment further on this phenomenon in chapter 13.

In order to state Theorem A we first need to extend the set Λ to allow for decoration of the groups by automorphisms. Let $X \in \Lambda$ and let $\mathrm{Aut}^o(X)$ be the largest subgroup of $\mathrm{Aut}(X)$ satisfying Hypothesis A, then $\mathrm{Aut}^o(X) = \mathrm{Aut}(X)$ or X is isomorphic to $L_3(p^n)$, $PSp_4(2^n)$, $G_2(3^n)$, in which case $\mathrm{Aut}^o(X)$ has index 2 in Aut X. We set

$$\Lambda^o = \{Y \mid X \le Y \le \mathrm{Aut}^o(X),\ X \in \Lambda\}.$$

Two more definitions and we are there. Suppose that G satisfies Hypothesis A. Set $Q_i = O_p(P_i)$ and $\bar{P}_i = P_i/Q_i$ and let

$$1 \le A_1 \le \dots \le A_3 \le Q_1 \le P_1 \quad \text{and} \quad 1 \le B_1 \le \dots \le B_6 \le Q_2 \le P_2$$

be normal series of P_1 and P_2, respectively.

(1) Suppose that $p = 2$ and the normal series can be chosen so that

(1a) $\bar{P}_1 \simeq Sz(2)$, $\bar{P}_2 \simeq SL_2(2)$,

(1b) $A_1 = Z(Q_1) = Z(P_1)$, $Z(P_2) = 1$,

(1c) A_2 and Q_1/A_2 are elementary abelian,

(1d) A_2/A_1 and A_3/A_2 are irreducible $\bar{P}_1$-modules of order 2^4, and $2 = |A_1| \ge |Q_1/A_3|$,

(1e) $B_6/B_4 \simeq C_4 \times C_4$, and B_3 is elementary abelian,

(1f) B_1, B_3/B_2, B_5/B_4, B_6/B_5 are natural $\bar{P}_2$-modules, and $2 = |B_2/B_1| = |B_4/B_3| \ge |Q_2/B_6|$.

Then G is of <u>type</u> ${}^2F_4(2)'$, if $|B| = 2^{11}$, and of <u>type</u> ${}^2F_4(2)$, if $|B| = 2^{12}$.

(2) G is of <u>type</u> F_3, if $p = 3$ and the normal series can be chosen so that $B_6 = Q_2$ and

(2a) $\bar{P}_1 \simeq \bar{P}_2 \simeq GL_2(3)$, $|B| = 4\cdot 3^{10}$,

(2b) A_2 and Q_1/A_2 are elementary abelian,

(2c) A_1, A_3/A_2, Q_1/A_3 are natural $\bar{P}_1$-modules, and A_2/A_1 is an irreducible non-faithful $\bar{P}_1$-module of order 3^3.

(2d) $Z(P_2) = Z(Q_2) = B_1$,

(2e) $Z(B_4) = B_3$,

(2f) B_1, B_3/B_2, B_5/B_4 are central $O^{3'}(\bar{P}_2)$-modules of order 3, and B_2/B_1, B_4/B_3, Q_2/B_5 are natural $\bar{P}_2$-modules.

Of course, as the notation suggests the groups ${}^2F_4(2)'$, ${}^2F_4(2)$, F_3 are type ${}^2F_4(2)'$, ${}^2F_4(2)$, F_3, respectively. It would be nice to be able to show that if G is of type ${}^2F_4(2)'$, ${}^2F_4(2)$ or F_3, then, indeed, G is parabolic isomorphic, or, better still, locally isomorphic to ${}^2F_4(2)'$, ${}^2F_4(2)$ or F_3, respectively. We will comment further on this at the end of the chapter.

With this definition at hand we are now able to state our main result.

<u>Theorem</u> A. <u>Suppose that</u> G <u>is a group with a weak</u> (B,N)-<u>pair of rank</u> 2. <u>Then one of the following holds</u>:

(a) G <u>is locally isomorphic to</u> X <u>for some</u> $X \in \Lambda^o$.

(b) G <u>is parabolic isomorphic to</u> $G_2(2)'$, J_2, $Aut(J_2)$, M_{12} <u>or</u> $Aut(M_{12})$.

(c) G <u>is of type</u> ${}^2F_4(2)'$, ${}^2F_4(2)$ <u>or</u> F_3.

As being locally isomorphic implies being parabolic isomorphic (see the definition at the end of chapter 2), Theorem A determines - apart from case (c) - P_1, P_2 and B up to isomorphism.

With Λ^o we have a fairly large class of examples satisfying Hypothesis A'. But it must be pointed out that these groups are not the only examples. In fact, very different groups can be parabolic isomorphic. We give some examples for pairs of groups (A,B) which are parabolic isomorphic and where $A \in \Lambda^o$: $(L_3(4), J_3)$, (J_2, J_3), $(\mathrm{Aut}(J_2), Ly)$, $(L_3(3), M_{12})$, $(\mathrm{Aut}(PSp_4(5)), .1)$. Here, J_3 is the third Janko group, Ly the Lyons group, and .1 the largest Conway group.

These examples are rather exotic. In fact, there is a general method for constructing infinitely many groups G_i all of which are parabolic isomorphic to some $X \in \Lambda$. Let $X \in \Lambda$ and let Y be the amalgamated product of X_1 and X_2 over $X_1 \cap X_2$. Clearly, Y has a normal subgroup U with $Y/U \simeq X$ and with $U \cap X_1 = 1 = U \cap X_2$. Hence U is free. It is a fact that U, as a free group, is residually finite; that is, for each $1 \neq u \in U$ there exists a normal subgroup N_u of finite index in U not containing u (see [11]). As N_u has finite index in Y, so does $\bigcap_{y \in Y} N_u^y = M_u$. So for each non-trivial element $u \in U$ we have constructed a normal subgroup M_u of finite index in Y not containing u. This construction clearly yields infinitely many finite groups each of which is parabolic isomorphic to X.

These groups are, in fact, more than parabolic isomorphic they are locally isomorphic.

The proof of Theorem A breaks up into three parts. We assume throughout the initial analysis that Hypothesis B' holds. Our comments below will always refer to this situation. In the first part we determine the parameter b as was illustrated in Example (3.10). We prove:

Theorem B. One of the following holds:

(a) $[Z_\alpha, Z_{\alpha'}] \neq 1$ and $b = 1$ or 2.

(b) $[Z_\alpha, Z_{\alpha'}] = 1$ and $b = 1, 3$ or 5.

Actually, a lot more is known at this stage of the proof, since the above cases (a) and (b) also impose restrictions on G_δ/Q_δ and p^{n_δ}. For example, if $b = 3$, then $p = 2$, and if $b = 5$, than $p^{n_\delta} = 3$.

In the next part of the proof, with b at hand, we determine the other parameters r and s and the structure of G_δ in certain exceptional cases as was illustrated in Example (3.9). We prove:

Theorem C. One of the following holds:

(a) $r = s-1$, and the stabilizers of paths of length s are p'-groups.

(b) G is parabolic isomorphic to $G_2(2)'$, M_{12}, $Aut(M_{12})$, or J_2.

(c) G is of type ${}^2F_4(2)'$ or F_3.

There seems to be no elegant approach to the identification of G_α and G_β (resp. P_1, P_2) in the exceptional cases (b) and (c). With regards to the groups in case (b) we describe most of the structure explicitly and leave some easy calculations to the reader. In the cases M_{12} and $Aut(M_{12})$ we determine generators and relations for G_α and G_β. The work here is not too extensive. Case (c) seems to need a laborious treatment with generators and relations and we only determine the weaker "type" as defined above. In the ${}^2F_4(2)'$-case Fan [5] has shown that G is, indeed, locally isomorphic to ${}^2F_4(2)'$. An analogous result is perhaps true in the F_3-case, however it is probably much harder to prove.

Nevertheless, in case (a) there is a uniform approach to the identification. This is the last part of the proof. We show that the condition $r = s-1$ implies the uniqueness condition and exchange condition for an appropriately chosen apartment or G is of type ${}^2F_4(2)$. Applying (3.7) we are able to conclude the following.

Theorem D. *Suppose that* $r = s-1$ *and the stabilizers of paths of length* s *are* p'*-groups, then* G *is locally isomorphic to* $L_3(p^n)$, $PSp_4(p^n)$, $G_2(p^n)$, $U_4(p^n)$, $U_5(p^n)$, ${}^3D_4(p^n)$, *or* ${}^2F_4(2^n)$ *or* G *is of type* ${}^2F_4(2)$.

Notice that we get here not only parabolic isomorphism but also local isomorphism. In the ${}^2F_4(2)$-case again Fan [5] has shown that G is locally isomorphic to ${}^2F_4(2)$.

It is then an easy task to pump-up the minimal configuration satisfying Hypothesis B' to the arbitrary one satisfying Hypothesis A'.

5. Properties of Chevalley groups of rank 1.

In this chapter we collect a number of properties of Chevalley groups of rank 1, in particular, their p-structure and some of their GF(p)-modules. The prime p throughout will denote the characteristic of the field of definition, hence $p = 2$ for $Sz(2^n)$, $p = 3$ for $Ree(3^n)$. Some of the results will be stated without proof as they are easily accessible in the literature, see, for example, Gorenstein [8] and Huppert [10]. Set

$$\Delta = \{L_2(p^n), SL_2(p^n), Sz(2^n), U_3(p^n), SU_3(p^n), Ree(3^n) \mid p \text{ prime}, n \in \mathbb{N}\}$$

and

$$\Delta^o = \{L_2(2), L_2(3), SL_2(3), Sz(2), U_3(2), SU_3(2), Ree(3)\} .$$

Note that none of the groups in Δ^o is perfect: $L_2(2) \simeq \Sigma_3$, $L_2(3) \simeq A_4$, $SL_2(3)/Z(SL_2(3)) \simeq A_4$, $Sz(2)$ is the Frobenius group of order 20, $U_3(2)$ and $SU_3(2)$ are groups of order 72, resp. 216, and are 3-closed, and $Ree(3) \simeq P\Gamma L_2(8)$ the extension of $L_2(8)$ by the field automorphism of GF(8) of order 3.

<u>Notation</u>. In the following $G \in \Delta$, $S \in Syl_p(G)$ and V is a faithful GF(p)G-module.

(5.1) <u>Let</u> K <u>be a complement to</u> S <u>in</u> $N_G(S)$.

(a) G <u>operates doubly transitive on</u> $\{N_G(T) \mid T \in Syl_p(G)\}$, <u>and</u> $N_G(S)$ <u>is a maximal subgroup of</u> G.

(b) If $G \notin \Delta^o$, then $G/Z(G)$ is simple and G is perfect.

(c) The Sylow p-subgroups of G are TI-groups, that is $T \cap S = 1$, $S \neq T \in Syl_p(G)$.

(d) G is generated by S and any p-element not in S.

(e) K is cyclic and $Z(G) \leq K$. If $G \notin \Delta^o$, then K acts irreducibly on $Z(S)$ and $S/\phi(S)$; $C_S(K) = 1$, $|N_G(K):K| = 2$, $C_G(K) = K$, and K normalizes exactly two Sylow p-subgroups.

(f) If $S = S_1 \times S_2$ with $1 \neq S_1 < S$, then $G = L_2(p^n)$ or $SL_2(p^n)$. If $S = S_1 \times S_2$ where S_1 is K-invariant, then $S_1 = 1$ or $S_1 = S$.

(g) If $x \in Aut(G)$ is a p'-element centralizing S, then $x = 1$.

Proof: (a) - (f) are well known. For (g) let $H = G\langle x\rangle$. If $G \in \Delta^o$, simple arguments which we leave to the reader yield the conclusion.

Suppose that $G \notin \Delta^o$. As $x \in N_H(S)$ and by [10,I, 18.2] S operates transitively on the set of complements to S in $N_G(S)$, there is a complement, without loss K, which is normalized by x. Now $[K,S,x] = 1 = [S,x,K]$ so by the Three-subgroup Lemma $[x,K,S] = 1$. Thus (e) implies $[x,K] = 1$ and so $[x,N_G(S)] = 1$. Again by (e) K normalizes exactly one further Sylow p-subgroup T, and so x normalizes T as well. As this holds for all complements to S in $N_G(S)$, (a) implies that x normalizes all Sylow p-subgroups and hence lies in the kernel M of the operation of H on $Syl_p(G)$. But by (b) $M \cap G = Z(G)$ and $G = G'$ and again the Three-subgroup Lemma yields $[x,G] = 1$ and $x = 1$. □

In the following table K is as in (5.1).

G	$\|G\|$	$\|K\|$	$\|Z(G)\|$
$SL_2(p^n)$	$p^n(p^n-1)(p^n+1)$	p^n-1	$(p^n-1,2)$
$Sz(2^n)$	$2^{2n}(2^n-1)(2^{2n}+1)$	2^n-1	1
$SU_3(p^n)$	$p^{3n}(p^{2n}-1)(p^{3n}+1)$	$p^{2n}-1$	$(p^n+1,3)$
$Ree(3^n)$	$3^{3n}(3^n-1)(3^{3n}+1)$	3^n-1	1

(5.2) *Let* $G = L_2(p^n)$ *or* $SL_2(p^n)$, K *a complement to* S *in* $N_G(S)$.

(a) S *is elementary abelian*.

(b) *Let* $p = 2$. *All involutions of* G *are conjugate*. *If* $U \leq S$, $|U| = 4$, *there exists* $g \in G$ *with* $\langle x, U^g\rangle = G$ *for* $1 \neq x \in U$.

(c) *Let* p *be odd*. G *has two conjugacy classes of elements of order* p. *If* $1 \neq x \in S$ *and* $p^n \neq 9$, *there exists* $g \in G$ *with* $\langle x, x^g\rangle = G$. *For* $p^n = 9$ $\langle x, x^g\rangle < G$ *for each* $g \in G$, *and there exists* $g \in G$ *with* $\langle x, x^g\rangle/Z(G) \simeq L_2(5)$, *and* $\langle x, x^g\rangle$ *is a maximal subgroup of* G. □

(5.3) *Let* $G \simeq Sz(2^n)$, $n > 1$, K *a complement to* S *in* $N_G(S)$.

(a) n *is odd and* 3 *does not divide the order of* G.

(b) $S/\phi(S)$ *and* $\phi(S)$ *are elementary abelian of order* 2^n.

(c) $Z(S) = \phi(S) = \Omega_1(S)$ *and* $C_S(k) = 1$ *for* $k \in K^\#$.

(d) *All involutions of* G *are conjugate*. *If* $U \leq Z(S)$, $|U| = 4$, $1 \neq x \in U$, *there exists* $g \in G$ *with* $\langle x, U^g\rangle = G$.

(e) *If* $S \neq T \in Syl_2(G)$, *then* $\langle Z(S), Z(T)\rangle = G$. □

(5.4) *Let* $G \simeq U_3(p^n)$ or $SU_3(p^n)$, K *a complement to* S *in* $N_G(S)$.

(a) $Z(S) = \phi(S)$ *is elementary abelian of order* p^n.

(b) If $p^n \neq 2$, then K operates irreducibly on $S/Z(S)$ and $Z(S)$.

(c) If $p = 2$, then $Z(S) = \Omega_1(S)$ and all involutions in G are conjugate. If p is odd, then $x^p = 1$ for every $x \in S$.

(d) If $S \neq T \in Syl_p(G)$, then $\langle Z(S),Z(T)\rangle \simeq SL_2(p^n)$.

(e) $K/Z(G)$ operates elementwise fixed-point-freely on $S/Z(S)$, and $|C_K(Z(S))/Z(G)| = (p^n+1)/(p^n+1,3)$.

(f) If $U \leq Z(S)$, $|U| \geq 4$, then there exist $g,h \in G$ with $\langle x,U^g,U^h\rangle = G$ for $1 \neq x \in U$. □

(5.5) Let $G \simeq Ree(3^n)$, K a complement to S in $N_G(S)$.

(a) n is odd, $|S| = 3^{3n}$.

(b) $\phi(S) = \Omega_1(S)$ has order 3^{2n} and $[S,\phi(S)] = Z(S)$ has order 3^n.

(c) K operates irreducibly on $S/\phi(S)$, $\phi(S)/Z(S)$ and $Z(S)$.

(d) G has elementary abelian Sylow 2-subgroups of order 8.

(e) If $n > 1$, then $G = \langle\phi(S),\phi(S)^g\rangle$ for $g \in G \setminus N_G(S)$. □

The data contained in (5.1) - (5.5) will, for the most part, be used without reference.

Next we state a classical number theoretic result.

(5.6) Let q be a prime, $m \in \mathbb{N}$. Then q^m-1 is divisible by a prime not dividing q^i-1 for $1 \leq i < m$, unless $q^m = 2^6$ or q^m is the square of a Mersenne prime.

Proof: See Zsigmondy [24] □

(5.7) _If_ G _is isomorphic to_, _respectively_, $SL_2(p^n)$, $L_2(p^n)$, $Sz(2^n)$, $SU_3(p^n)$, $U_3(p^n)$, _then the_ GF(p)-_dimension of the smallest faithful_ GF(p)-_module is at least_, _respectively_, 2n, 2n, 4n, 6n, 6n.

Note: For $L_2(p^n)$ ($\neq SL_2(p^n)$) and $U_3(p^n)$ ($\neq SU_3(p^n)$) there is no module of this smallest dimension. But we will not need this result.

Proof: We will show that the order of G does not divide the order of $GL_m(p)$ if m is smaller than the dimensions given. We use (5.6). Suppose first that G is isomorphic to $L_2(p^n)$ or $SL_2(p^n)$. The order of G is divisible by p^n+1. Except for the exceptional cases of (5.6) there is a prime r dividing $p^{2n}-1$ but not dividing p^i-1 for $1 \leq i < 2n$. In particular, r is neither 2 nor divides p^n-1. Hence r divides p^n+1. A glance at the order of $GL_m(p)$ and we are done. If $G \simeq SL_2(2^3)$, then G has cyclic Sylow 3-subgroups of order 9. But $GL_m(2)$ has elementary abelian Sylow 3-subgroups for $m \leq 5$. If G is isomorphic to $L_2(p)$ or $SL_2(p)$, the assertion is obvious.

In the case $G \simeq Sz(2^n)$ consider $2^{4n}-1$ and for $G \simeq U_3(p^n)$ or $SU_3(p^n)$ consider $p^{6n}-1$. The arguments, being just like those above, are left to the reader. □

We will write all GF(p)G-modules multiplicatively and use the exponential notation for the operation of G on V, since for us modules will only occur as sections of groups and the operation is induced by conjugation.

Definition. Let X be a group and V a $GF(p)X$-module. A subgroup $A \leq X$ operates quadratically on V, if $[V,A,A] = 1$.

We formulate the next lemma for a general group X.

(5.8) Suppose X is a finite group and V is a faithful $GF(p)X$-module. Let $A \leq X$ with $[V,A,A] = 1$.

(a) If $V = \langle C_V(A^g) \mid g \in X\rangle$, then $V = [V,X]C_V(A)$.

(b) A is an elementary abelian p-subgroup.

(c) For $a \in A$, $|V/C_V(a)| = |[V,a]|$.

(d) For $g \in X$ and $1 \neq t \in A$ set $H = \langle t,A^g\rangle$ and $\tilde{V} = V/C_V(H)$. If $O_p(H) = 1$, then $[\tilde{V},H] = [\tilde{V},t] \times [\tilde{V},A^g]$ and $[\tilde{V},A^g] \cap C_{\tilde{V}}(t) = 1$.

Proof: $[V,X]C_V(A)$ is a X-submodule of V containing $C_V(A)$ and (a) holds.

By the Three-subgroup Lemma $[A,A,V] = 1$. Hence, A' centralizes V and A is abelian. For $v \in V$, $a \in A$, $[v,a^2] = [v,a][v,a]^a$. From $[V,A,A] = 1$ we get $[v,a^2] = [v,a]^2$ and, by induction, $[v,a^p] = [v,a]^p = 1$. Thus $a^p = 1$ for $a \in A$ and (b) holds.

The map $v \longrightarrow [v,a]$ is $GF(p)$-linear. Hence (c) follows.

Note that $C_H(\tilde{V})$ operates quadratically on V, i.e. $[V,C_H(\tilde{V}),C_H(\tilde{V})] = 1$. Thus by (b), $C_H(\tilde{V}) = 1 = C_{\tilde{V}}(t) \cap C_{\tilde{V}}(A^g)$. Now (d) follows, since obviously $[\tilde{V},A] \leq C_{\tilde{V}}(A)$ and $[\tilde{V},H] = [\tilde{V},t][\tilde{V},A^g]$. □

We now return to having $G \in \Delta$ and $S \in Syl_p(G)$, and V a faithful $GF(p)G$-module.

(5.9) Let $1 \neq A \leq S$ so that $[V,A,A] = 1$. Then:

(a) G is not isomorphic to $L_2(p^n)$, p odd, or $Ree(3^n)$.

(b) $A \leq \Omega_1(Z(S))$.

Proof: $L_2(p^n)$ has for p odd dihedral Sylow 2-subgroups and $Ree(3^n)$ has abelian Sylow 2-subgroup. Now [8, 3.8.3] yields (a).

From (5.8)(b) we need only show that $A \leq Z(S)$. If G is one of $SL_2(p^n)$, $Sz(2^n)$, $U_3(2^n)$, or $SU_3(2^n)$ the result follows from the structure of S (see (5.2) - (5.4)). Hence we may assume that $G \simeq U_3(p^n)$ or $SU_3(p^n)$ with p odd. Let $1 \neq x \in A$ and choose $g \in G$ so that $M = \langle x,x^g \rangle$ is not a p-group. By [8, 3.8.1] M is isomorphic to $SL_2(p^m)$ for $m \in \mathbb{N}$ or $SL_2(5)$. (This is not precisely the result stated there, but this is contained in the proof). In any case, there is an involution $t \in Z(M)$ and, as $|Z(G)| \mid 3$, $t \notin Z(G)$. Since t normalizes a p-group (5.1)(c) implies that $t \in N_G(S)$. Now the structure of $N_G(S)$ (see (5.4)(e)) shows $x \in Z(S)$. Hence $A \leq Z(S)$ and we are done. □

(5.10) Let $A \leq S$, and suppose that $|A| \geq 3$, and $[V,A,A] = 1$. Then for $1 \neq t \in A$:

(a) If $G \simeq SL_2(p^n)$, then $|V/C_V(t)| \geq p^n$.

(b) If $G \simeq Sz(2^n)$, $U_3(p^n)$ or $SU_3(p^n)$, then $|V/C_V(t)| \geq p^{2n}$.

(c) If $G \simeq SL_2(p^n)$ or $Sz(2^n)$ and $V = \langle C_V(A)^G \rangle$, then $C_V(A) = C_V(t)$.

Proof: Let V be a counter example of minimal dimension. By the quadratic operation of A we have $V = [V,G]$ and $V = \langle C_V(A)^G \rangle$. Choose $g \in G$ and $L = \langle t,A^g \rangle$ so that L has maximal order.

Assume first that $L = G$. According to (5.8)(d) we have $V = C_V(G)[V,t][V,A^g]$ and $C_V(t) = C_V(G)[V,t] = C_V(A)$. In addition, by (5.8)(c) $|V/C_V(G)| \leq |[V,t]|^2 = |V/C_V(t)|^2$ and (5.7) implies $|V/C_V(t)| \geq q$ if $G \simeq SL_2(q)$ and $|V/C_V(t)| \geq q^2$ if $G \simeq Sz(q)$.

Assume next that $L \simeq SL_2(5)$, $|A| = 3$ and $G \simeq SL_2(9)$. Then $Z(L) = Z(G)$, and by (5.9)(a) G centralizes $C_V(Z(L))$. It follows that $C_V(G) = C_V(L)$, and the above argument applied to L and $[V,L]$ yields $C_V(A) = C_V(t)$. Since $SL_2(5)$ cannot operate non-trivially on a 2-dimensional $GF(3)$-module, we also get $|V/C_V(t)| \geq 9$. We conclude (a) and (c), and (b) if $G \simeq Sz(2^n)$.

According to (5.2) - (5.4) we may assume now that $G \simeq U_3(q)$ or $SU_3(q)$ and $L \simeq SL_2(q)$ or $L \simeq SL_2(5)$ and $q = 9$. If $L \simeq SL_2(5)$ and $q = 9$, then $L \leq L_o \simeq SL_2(9)$ and $Z(L) = Z(L_o)$, and by the quadratic operation of A we have $[V,L] = [V,Z(L)] = [V,L_o]$. Set $L = L_o$ in the other cases and $V_1 = [V,L_o]$ and $V_o = C_V(A)C_V(t)$. As we have seen $V_1 \leq V_o$ and L_o normalizes V_o. Note that $A \leq Z(S) \leq L_o$. Now (c) gives $C_{V_o}(A) = C_{V_o}(a)$ for $1 \neq a \in A$. Since we may assume that $|V_o/C_{V_o}(t)| < q^2$, we get $|V_o/C_{V_o}(L_o)| < q^4$.

If q is odd, then $V_1 = [V,Z(L_o)]$ and $C_{V_1}(L_o) = 1$. It follows that $|[V,A]| = |C_{V_o}(t)| < q^2$, and (5.4) and (5.7) yield a contradiction.

Assume now that q is even. Since $|A| \geq 4$ we have $q \neq 2$. Hence, again by (5.4) there exist non-trivial subgroups K^+ and K^- in $N_G(S)$ so that

(i) $|K^+| = q+1$ or $q+1/3$, $|K^-| = q-1$,

(ii) $[K^+,L_o] = 1$, $K^- \leq L_o$,

(iii) $[S,k] = S$ for $1 \neq k \in K^+$.

If $[V_0,k] = 1$ for some $1 \neq k \in K^+$, then the Three-subgroup-Lemma applied to $[V,k,A]$ and $[S,k,V]$, successively, yields $[V,k] \leq C_V(A)$ and then $[S,V] \leq C_V(A)$. It follows that S normalizes V_0, and since by (5.1) $G = \langle S,A^g \rangle$, we get $V = V_0$. But now (5.4)(f) and $|V/C_V(A)| < q^2$ imply $|V/C_V(G)| < q^6$ contradicting (5.7).

If $[V_0,k] \neq 1$ for every $1 \neq k \in K^+$, then K^+ operates fixed-point-freely on $V_0/C_{V_0}(L_0)$. Now an application of (5.6) yields $|K^+| = 3$ or 9, $q = 2^3$ and $|V_0/C_{V_0}(L_0)| = 2^8$, since K^+ operates on $C_{V_0}(A)/C_{V_0}(L_0)$. But then $|A|^2 > |Z(S)|$ and $C_{V_0}(t) = C_{V_0}(Z(S))$, in particular K^+K^- normalizes $C_{V_0}(t)$. Since $|K^-| = 7$ we get $|C_{V_0}(t) \cap C_{V_0}(K^-)/C_{V_0}(L_0)| = 2$ and K^+ does not operate fixed-point-freely on $V_0/C_{V_0}(L_0)$, a contradiction. □

(5.11) Let $G \simeq SL_2(q)$, q even, or $Sz(q)$, and $Z = \Omega_1(Z(S))$. Suppose that $V = C_V(Z) \times C_V(Z^g)$ for $g \in G \setminus N_G(S)$. Then the non-trivial elements of odd order in G operate fixed-point-freely on V.

Proof: Note that by (5.2) and (5.3) $O^2(G) \leq \langle Z^g,x \rangle$ for $1 \neq x \in Z$. It follows that $C_V(Z) = C_V(x)$. Note further that every element $y \neq 1$ of odd order in G is inverted by an involution (see [8] and [20]). So we may assume that $y^x = y^{-1}$. Then $y^2 = xx^y$ and $C_V(x) \cap C_V(x^y) = 1$ since $y \notin N_G(S)$. This yields $C_V(y^2) = C_V(y) = 1$. □

Definition. Let $G = SL_2(p^n)$, $S \in Syl_2(G)$ and V be a $GF(p)G$-module. Then V is a natural $SL_2(p^n)$-module, if

(i) $|V| = p^{2n}$,

(ii) $[V,G'] \neq 1$ and $[V,S,S] = 1$.

(In fact, V is the natural 2-dimensional $GF(p^n)G$-module considered as $GF(p)G$-module, but we will not need this result.)

According to (5.7), (5.11) and the quadratic action of S we have the following properties of a natural module which we will use in the following:

(i) V is irreducible.

(ii) $|C_V(S)| = |C_V(t)| = p^n$ for $1 \neq t \in S$.

(iii) $V = C_V(S) \times C_V(S^g)$, $g \in G \setminus N_G(S)$.

(iv) If p is even, then the non-trivial elements of V are conjugate under G.

(5.12) Let $A \leq S$ and suppose that $[V,A,A] = 1$ and $C_V(O^p(G)) = 1$. Then

(a) $|A| \leq |V/C_V(A)|$,

(b) if $|A|^2 > |V/C_V(A)|$, then $G \simeq SL_2(p^n)$ and V is a natural $SL_2(p^n)$-module.

Proof: Note first that from (5.9)(a) G is not isomorphic to either $L_2(p^n)$, p odd, or $Ree(3^n)$, and from (5.9)(b) we know that $|A|$ is bounded by $|\Omega_1(Z(S))|$. The structure of S implies $|A| \leq p^n$ where $GF(p^n)$ is the field of definition for G. Let k be the smallest number of conjugates of A needed to have $O^p(G) \leq \langle A^{g_1},\ldots,A^{g_k}\rangle$. From (5.2) - (5.4), $k \leq 3$ if $G \simeq SL_2(p^n)$ or $Sz(2^n)$, and $k \leq 4$ if $G \simeq U_3(p^n)$ or $SU_3(p^n)$.

Note that $C_V(O^p(G)) = 1$ and thus $|V| \leq |V/C_V(A)|^k$. If $|A| = p$, then clearly (a) holds. If $|A| > p$, then $k = 2$ or 3, and (5.7) implies (a).

It remains to prove (b). So assume that $|A|^2 > |V/C_V(A)|$. Then

$$|V| \leq p^{-k}|A|^{2k} < p^{2nk}.$$

Suppose first that $|A| = p$. Then $|V| \leq p^k$, and (b) follows if $k = 2$. Let $H = \langle A, A^x \rangle$, $x \in G$, be of maximal order. Then $|V/C_V(H)| \leq p^2$ and by (5.2) $H \simeq SL_2(p)$ and $G \simeq U_3(p)$ or $SU_3(p)$; in particular $k = 3$ and $|V| \leq p^3$. But this contradicts (5.7).

We can now suppose that $|A| > p$. From (5.2) - (5.4) we conclude that $k = 2$ if $G \simeq SL_2(p^n)$ or $Sz(2^n)$ and $k = 3$ if $G \simeq U_3(p^n)$ or $SU_3(p^n)$. Thus, by (5.7), we have $k = 2$ and $G \simeq SL_2(p^n)$. In particular, $|V| < |A|^4$ and again (5.7) yields $|A| > p^{n/2}$.

For any $a \in A^{\#}$ there exists $g \in G$ so that $G = \langle a, A^g \rangle$ (see (5.2)). Hence, by (5.8)(d) $V = C_V(a) \times C_V(A^g)$, and as $C_V(A) \leq C_V(a)$ we get $C_V(A) = C_V(a)$. The operation of $N_G(S)$ on S and $|A| > p^{n/2}$ now yields $C_V(A) = C_V(s) = C_V(S)$ for $s \in S^{\#}$. Since $[A,V,S] = [S,A,V] = 1$ the Three-subgroup-Lemma gives $[V,S,S] = 1$. Thus, we may assume $A = S$.

We now treat the cases $p = 2$ and p odd separately. Suppose first that $p = 2$. Then by (5.11) every non-trivial element of odd order operates fixed-point-freely on V; in particular the non-trivial elements in K where K is a complement to S in $N_G(S)$.

Since $|K| = p^n-1$, we have for every non-trivial K-submodule U in V that $p^n-1 \mid |U|$. From $|V| < |A|^4 \leq p^{4n}$ we get $|C_V(S)| < p^{2n}$ and thus $|C_V(S)| = p^n$, since $C_V(S)$ is a K-submodule. Hence, $|V| = p^{2n}$ and the result follows.

Next suppose that p is odd. We may clearly suppose that V is irreducible. Set $L = GF(p)$, $M = GF(p^n)$ so that M is a splitting field for G. Put $\tilde{V} = V \otimes_L M$, then $\tilde{V} \simeq \bigoplus_{i=1}^{k} W^{\sigma_i}$ where W is a faithful irreducible MG-module and $\sigma_1,\ldots,\sigma_k \in Gal(M/L)$.

We claim $\dim_M(W) = 2$. By [8, 2.8.4], if $p^n \neq 3^2$, the matrices

$$x_1 = \begin{pmatrix} 1 & 0 \\ 1 & 1 \end{pmatrix}, \qquad x_2 = \begin{pmatrix} 1 & \lambda \\ 0 & 1 \end{pmatrix}$$

generate G where λ is a generator of M over L. Now following the proof of [8, 3.8.1] step by step we see that $\dim_M(W) = 2$ as claimed. We conclude that $\dim_L V = \dim_M \tilde{V} \leq 2n$. By (5.7) $\dim_L V = 2n$ and the result holds.

Suppose that $p^n = 3^2$. If $|V| \neq 3^4$, then $|V| = 3^6$, $|C_V(S)| = 3^3$ and $|K| = 8$. Note that $|Z(G)| = 2$ and $Z(G) \leq K$. Since V is irreducible, Z(G) operates fixed-point-freely on V, and since K is cyclic, every non-trivial element in K operates fixed-point-freely on V. Now we get a contradiction as above in the case $p = 2$. □

(5.13) <u>Let</u> X <u>be a finite group and</u> $Z \leq N \leq X$ <u>a normal series of</u> X <u>so that</u>

(i) $X/N \simeq SL_2(q)$, q <u>even, and</u> N/Z <u>is a natural</u> $SL_2(q)$-<u>module</u>.

(ii) $Z \leq \Omega_1(Z(X))$.

<u>Then either</u> $q = 2$ <u>and</u> $\Omega_1(N) \leq Z$, <u>or</u> N <u>is elementary abelian</u>.

<u>Proof</u>: Suppose N to be not elementary abelian. Then we may assume that $|Z| = 2$. Since N/Z is a natural module, the non-trivial elements in N/Z are conjugate under X. It follows that there are no involutions in $N \setminus Z$. Hence, $N \simeq Q_8$ and $q = 2$. □

<u>Definition</u>: Let $G = SL_2(4)$ and $S \in Syl_2(G)$. Then V is an <u>orthogonal</u> $SL_2(4)$-<u>module</u>, if $|V| = 4^2$ and $[V,S,S] \neq 1$.

(5.14) <u>Let</u> X <u>be a finite group and</u> N <u>be a normal subgroup of</u> X. <u>Suppose that</u>

(i) N <u>is extra special of order</u> 2^5 <u>and</u> $C_X(N) \leq N$,

(ii) $X/N \simeq SL_2(4)$.

<u>Then</u> $N/Z(N)$ <u>is an orthogonal</u> $SL_2(4)$-<u>module</u>.

<u>Proof</u>: Set $V = N/Z(N)$, and pick $S \in Syl_2(X)$. If $[V,S,S] = 1$, then V is a natural module which contradicts (5.13). Hence, V is an orthogonal $SL_2(4)$-module. □

(5.15) <u>Let</u> $X = HN$ <u>where</u> $H \simeq SL_2(4)$ <u>and</u> N <u>is an orthogonal</u> $SL_2(4)$-<u>module for</u> H. <u>Let</u> $S \in Syl_2(X)$ <u>and</u> K <u>a complement to</u> S <u>in</u> $N_X(S)$. <u>Then there are exactly three</u> K-<u>invariant fours-groups</u> A <u>in</u> S with $[A,K] = A$.

<u>Proof</u>: Set $S_o = [S,K]$. Since N is an orthogonal module, it is easy to check that S_o is extra special of order 2^5, and K operates fixed-point-freely on $S_o/Z(S_o)$. Now the result is easy to check. □

(5.16) <u>Let</u> X <u>be as in</u> (5.14), $S \in Syl_2(X)$ <u>and</u> K <u>a complement to</u> S <u>in</u> $N_X(S)$. <u>Suppose there are two</u> K-<u>invariant complements to</u> N <u>in</u> S. <u>Then</u> $X = HN$, $H \simeq SL_2(4)$ <u>and</u> $N \simeq Q_8 * D_8$.

<u>Proof</u>: Let A_1 and A_2 be the two given K-invariant complements to N in S and set $\bar{X} = X/Z(N)$. By (5.15) $\bar{A}_1$ and $\bar{A}_2$ are the only K-invariant fours-groups in S outside $\bar{N}$. In addition, Gaschütz's Theorem [10] gives us a complement $\bar{H} \simeq SL_2(4)$ to $\bar{N}$. We may assume that $\bar{K}\bar{A}_1 \leq \bar{H}$. Let H be the preimage of $\bar{H}$. Then $H \simeq C_2 \times SL_2(4)$ and the assertion follows, since the isomorphism type of N is uniquely determined by the operation of H. □

(5.17) [9, 8.2]. <u>Let</u> X <u>be a finite group so that</u> $X/O_2(X) \simeq L_2(2^n)$, $n > 1$. <u>Suppose that an element of</u> X <u>of order</u> 3 <u>acts fixed-point-freely on</u> $O_2(X)$. <u>Then</u> $O_2(X)$ <u>is elementary abelian and a direct product of natural</u> $X/O_2(X)$<u>-modules</u>. □

(5.18) [12]. <u>Let</u> X <u>be a finite group so that</u> $X/O_2(X) \simeq Sz(2^n)$, $n > 1$. <u>Suppose that an element of</u> X <u>of order</u> 5 <u>acts fixed-point-freely on</u> $O_2(X)$. <u>Then</u> $O_2(X)$ <u>is elementary abelian</u>. □

6. Interlude: Notations and preliminary results.

(6.0) Hypothesis. In this chapter Hypothesis B' holds.

For $\delta \in \Gamma$ the subgroups $Q_\delta, Z_\delta, V_\delta, V_\delta^{(i)}$ and the sets $\Delta(\delta)$, $\Delta^{(i)}(\delta)$ and the parameters b_δ, b, r and s are defined as in chapter 3. In addition, we adopt the special choice of α, β, α' and γ ; in particular $\beta = \alpha+1 = \alpha'-(b-1)$ and $G_\alpha \cap G_\beta = B$.

Since B is p-closed, by [8,6.2.1] there exists a complement to $O_p(B)$ in B. We define:

$$S = O_p(B),$$
$$K \quad \text{a complement to } S \text{ in } B,$$
$$L_\delta = SG_\delta^*, \quad \delta = \alpha,\beta,$$
$$K_\delta = K \cap L_\delta, \quad \delta = \alpha,\beta,$$
$$L_\mu = L_\delta^g, \quad g \in G \text{ with } \delta^g = \mu,$$
$$q_\mu = |\Omega_1(Z(T))| \quad \text{for } T \in Syl_p(L_\mu/Q_\mu).$$

Obviously, the definition of L_μ is independent of the choice of g. Note that G_μ/L_μ is a p'-group and $K = K_\alpha K_\beta$. With one exception this notation will be fixed for the remainder of this book, the exception being V_β in chapter 11.

(6.1) For $\delta = \alpha,\beta$ the subgroup L_δ fulfils Hypothesis B' in place of G_δ^*.

Proof: By definition L_δ/G^*_δ is a p-group. If $G^*_\delta/Q_\delta \simeq SL_2(2)$, D_{10}, $Sz(2)$, $L_2(3)$, $SL_2(3)$ or $Ree(3)'$ the assertion is easy to check. In the other cases $K_\delta \not\leq G^{(1)}_\delta$ (see chapter 5). Since $[K_\delta,S] \leq S \cap G^*_\delta$ we have $S = C_S(K_\delta)(S \cap G^*_\delta)$ and an application of (5.2) - (5.5) yields $C_S(K_\delta) \leq G^*_\delta C_S(G^*_\delta/Q_\delta)$. Now the transitivity of G^*_δ on $\Delta(\delta)$ implies $C_S(G^*_\delta/Q_\delta) = Q_\delta$. □

Remark: The proof of (6.1) shows that $L_\delta = G^*_\delta$ or $G^*_\delta/Q_\delta \simeq D_{10}$ or $Ree(3)'$ and $L_\delta/Q_\delta \simeq Sz(2)$ or $Ree(3)$, respectively.

(6.2) Let $\delta \in \Gamma$ and N be a normal subgroup of G_δ in L_δ which is not p-closed. Then

(a) $O^p(L_\delta) \leq N$,

(b) N is transitive on $\Delta(\delta)$,

(c) $Q_\delta \in Syl_p(C_{L_\delta}(Z_\delta))$ or $Z_\delta = \Omega_1(Z(L_\delta))$,

(d) $Q_\delta \in Syl_p(G^{(1)}_\delta)$.

Proof: Note that L_δ is not p-closed. Assertions (a) and (b) follow from (6.1) and the definition of L_δ.

If $Q_\delta \notin Syl_p(C_{L_\delta}(Z_\delta))$ then $C_{L_\delta}(Z_\delta)$ is not p-closed, and we may choose $N = C_{L_\delta}(Z_\delta)$. By (a) we get $L_\delta = NT$, $T \in Syl_p(L_\delta)$, and $Z_\delta = \langle \Omega_1(Z(T))^g \mid g \in NT\rangle = \Omega_1(Z(T))$, and (c) follows.

Since $G^{(1)}_\delta$ is not transitive on $\Delta(\delta)$, we conclude from (b) that $G^{(1)}_\delta$ is p-closed. Now (d) follows from (3.8)(b). □

For any p-group P we define $\mathcal{A}(P)$ to be the set of elementary abelian subgroups of P of maximal order, and $J(P) = \langle A \mid A \in \mathcal{A}(P)\rangle$.

In the following lemmata we will treat a special pushing up situation which arises in the proof of Theorem A. It is remarkable that for odd primes p an elementary argument eliminates the situation, while for $p = 2$ Baumann's pushing up result [1] is needed.

(6.3) [1]. <u>Let</u> L <u>be a finite group</u>, $P \in Syl_2(L)$ <u>and</u> $V = \langle \Omega_1(Z(P))^L\rangle$. <u>Suppose that</u>

(i) $L/O_2(L) \simeq SL_2(2^n)$,

(ii) $C_L(O_2(L)) \leq O_2(L)$,

(iii) <u>no non-trivial characteristic subgroup of</u> P <u>is normal in</u> L.

<u>Then</u> $[O_2(L),L] \leq V$, <u>and</u> $V/V \cap Z(L)$ <u>is a natural</u> $SL_2(2^n)$-<u>module</u>.

□

(6.4) <u>Let</u> δ,λ <u>be adjacent in</u> Γ <u>and</u> $L = \langle Q_\delta^{G_\lambda}\rangle$. <u>Suppose that</u> $Q_\delta \cap Q_\lambda$ <u>is normal in</u> L_λ. <u>Then</u>

(a) $Q_\delta \in Syl_p(L)$ <u>and</u> $Z_\lambda \leq C_L(O_p(L)) \leq O_p(L)$,

(b) <u>no non-trivial characteristic subgroup of</u> Q_δ <u>is normal in</u> L,

(c) $L/O_p(L) \simeq SL_2(q_\lambda)$ <u>and</u> $Q_\delta Q_\lambda \in Syl_p(L_\lambda) \cap Syl_p(L_\delta)$.

<u>Proof</u>: Since L is normal in G_λ, we get $Z_\lambda \leq C_L(O_p(L)) \leq O_p(L)$ from Hypothesis B' and the definition of Z_λ.

By (3.2)(c) $Q_\delta \not\leq Q_\lambda$ and by (6.2)(b) L is transitive on $\Delta(\delta)$. Now (b) follows again from (3.2)(c).

We next prove $L/O_p(L) \simeq SL_2(p^n)$. Pick $A \in \mathcal{A}(Q_\delta)$ and set $V = \langle Z_\delta^L\rangle$. By (b) we may assume that $A \not\leq O_p(L)$ and by the Thompson Replacement Theorem [8, 8.2.5] that $[V,A,A] = 1$. Since by the maximality of A

$$|A/C_A(V)| \geq |V/C_V(A)|$$

the claim follows from (5.12).

Since $LQ_\lambda = L_\lambda$ and $Q_\delta O_p(L) \in Syl_p(L)$, it remains to prove that $O_p(L) \leq Q_\delta$. Clearly $[O_p(L),Q_\delta] \leq Q_\delta \cap Q_\lambda$, and $L/Q_\delta \cap Q_\lambda$ is a central extension of $SL_2(p^n)$ by a p-group. Now [10,I, 17.4] shows that Q_δ covers a Sylow p-subgroup of $L/Q_\delta \cap Q_\lambda$ and $Q_\delta \in Syl_p(L)$. □

(6.5) Let δ,λ be adjacent in Γ and $L = \langle Q_\delta^{G_\lambda}\rangle$. Then

(a) $Q_\delta \cap Q_\lambda$ is not normal in G_λ,

(b) $Q_\delta \notin Syl_p(L)$.

Proof: If $Q_\delta \in Syl_p(L)$, then $Q_\delta \cap O_p(L) = Q_\delta \cap Q_\lambda$ is normal in G_λ. Thus, we may assume by way of contradiction that $Q_\delta \cap Q_\lambda$ is normal in G_λ.

Assume first that $p = 2$ and set $V = \langle \Omega_1(Z(Q_\delta))^L\rangle$. Then (6.3) and (6.4) imply that $L/O_2(L) \simeq SL_2(q_\lambda)$, $[O_2(L),L] \leq V$, and $V/V \cap Z(L)$ is a natural $SL_2(q_\lambda)$-module. It follows that $Z_\lambda \leq Z(L)$ or $Z_\lambda(V \cap Z(L)) = V$.

Suppose that $Z_\lambda \leq Z(L)$. Let $T \in Syl_2(G_\lambda)$ and note that $[Q_\lambda,L] \leq O_2(L)$. If $q_\lambda = 2$, then $V = C_V(O^2(L)) \times [V,O^2(L)]$ and $[V,O^2(L)]$ is normal in T. Hence, $Z(T) \cap [V,O^2(L)] \neq 1$ and $Z_\lambda \nleq Z(L)$, a contradiction.

If $q_\lambda > 2$, let $\tilde{K}$ be a complement to $T \cap L$ in $N_L(T \cap L)$.

Then $T = C_T(\tilde{K})(T \cap L)$ and $V = C_V(\tilde{K}) \times [V,\tilde{K}]$, and again $Z(T) \cap [V,\tilde{K}] \neq 1$, contradicting $Z_\lambda \leq Z(L)$.

Suppose now that $Z_\lambda \nleq Z(L)$. Then we have $[O_2(L),O^2(L)] \leq Z_\lambda$. If $V_\delta \leq O_2(L)$, then V_δ is normal in L and G_δ, contradicting (3.2)(c). Thus, there exists $\mu \in \Delta(\delta)$ so that $Z_\mu \leq L$ but $Z_\mu \nleq O_2(L)$, i.e. $b = 2$. We first show that there is an element x of odd order in G_δ with $\lambda^x = \mu$. Set $\bar{L}_\delta = L_\delta/Q_\delta$. The Sylow 2-subgroups of $\bar{L}_\delta$ are TI-groups. Choose $t \in O_2(G_\lambda \cap G_\delta)$ and $t' \in O_2(G_\mu \cap G_\delta)$. Then $\bar{t}\bar{t}'$ has odd order, and we may pick $x \in \langle tt' \rangle$.

It is easy to check that $Q_\delta = Z_\lambda^x Z_\lambda D$ where $D = O_2(L) \cap O_2(L)^x$; in particular, $O_2(L) = DZ_\lambda$ and $O_2(L^x) = DZ_\lambda^x$. It follows that

$$\phi(O_2(L)) = \phi(D) = \phi(O_2(L^x))$$

and $\phi(D) = \phi(D)^x$. Hence, x operates on $Q_\delta/\phi(D)$. But $Q_\delta/\phi(D)$ contains exactly two maximal elementary abelian subgroups, namely $O_2(L)/\phi(D)$ and $O_2(L^x)/\phi(D)$. Since x has odd order, it has to normalize both and $Z_\lambda^x = Z_\mu \leq O_2(L)$, a contradiction.

Assume now that p is odd. Since $L/O_p(L) \simeq SL_2(p^n)$ there exists an involution t in L whose image generates $Z(L/O_p(L))$. It follows that $C_L(t)$ operates transitively on $\Delta(\lambda)$ and $Q_\lambda = C_{Q_\lambda}(t)O_p(L)$. On the other hand, by (6.4)(c) $C_{Q_\lambda}(t)Q_\delta$ is a Sylow p-subgroup of L_δ. Since L_δ is not p-closed by (6.1), we get from (5.1)(g) that $[L_\delta,t] \leq Q_\delta$. Hence, $C_{L_\delta}(t)$ operates transitively on $\Delta(\delta)$, a contradiction to (3.2)(c). □

(6.6) Let δ,λ be adjacent in Γ and $T \in Syl_p(G_\delta \cap G_\lambda)$. Suppose that $Z_\lambda Z_\delta$ is normal in G_λ. Then $Z_\delta = \Omega_1(Z(T))$ and $Z_\delta \leq Z_\lambda$.

Proof: Note that $T \in Syl_p(G_\lambda) \cap Syl_p(G_\delta)$. Thus $Z_\delta = \Omega_1(Z(T))$ implies $Z_\delta \leq Z_\lambda$. If $Z_\delta \neq \Omega_1(Z(T))$, then (6.2)(c) yields $Q_\delta \in Syl_p(C_{L_\delta}(Z_\delta))$. It follows that $Q_\delta \cap Q_\lambda = C_{Q_\lambda}(Z_\delta Z_\lambda)$ is normal in G_λ, and (6.5) implies a contradiction. □

(6.7) Let $\tilde{\gamma}$ be a path of length s (s as in chapter 3) containing (α,β). Then $B = (B \cap G_{\tilde{\gamma}})S$.

Proof: Since $|\Delta(\delta)| - 1$ is a p-power, the result follows from (3.4). □

(6.8) Suppose that $L_\delta/Q_\delta \simeq L_2(q_\delta)$ or $SL_2(q_\delta)$ for $\delta \in \Gamma$. Let $\hat{\gamma} = (\delta_o,\ldots,\delta_s)$ be a path of length s so that $G_{\hat{\gamma}}$ is not transitive on $\Delta(\delta_s) \setminus \{\delta_{s-1}\}$. Then

(a) $O_p(G_{\hat{\gamma}}) \leq G^{(1)}_{\delta_s}$,

(b) $s \geq 4$.

Proof: We may assume (after conjugation) that $\{\delta_o,\delta_1\} = \{\alpha,\beta\}$, and in addition by (6.7) that $K \leq G_{\hat{\gamma}}$. If $O_p(G_{\hat{\gamma}}) \not\leq G^{(1)}_{\delta_s}$, then the operation of K on $O_p(G_{\hat{\gamma}})Q_{\delta_s}/Q_{\delta_s}$ (see (5.1)) implies that $G_{\hat{\gamma}}$ is transitive on $\Delta(\delta_s) \setminus \{\delta_{s-1}\}$, a contradiction.

Assume now that $s = 2$. Then by (a) $Q_{\delta_1} \leq Q_{\delta_2}$; and again the operation of K yields $Q_{\delta_2} \in Syl_p(L_{\delta_2})$, but $G^*_{\delta_2}$ is not p-closed.

Assume next that $s = 3$. Then, again by (a), $Q_{\delta_1} \cap Q_{\delta_2} \leq Q_{\delta_3}$ and thus $Q_{\delta_1} \cap Q_{\delta_2} = Q_{\delta_3} \cap Q_{\delta_2}$ It follows that $Q_{\delta_1} \cap Q_{\delta_2}$ is normal in L_{δ_2}, contradicting (6.5). □

(6.9) <u>Let</u> $L_\delta/Q_\delta \simeq SL_2(q_\delta)$, q_δ <u>odd, for</u> $\delta \in \Gamma$. <u>Suppose that</u> $\tilde{\gamma} = (\alpha,\beta,\delta_2,\delta_3,\delta_4)$ <u>is a path of length</u> 4 <u>normalized by</u> K. <u>Set</u> $\langle t_\delta\rangle = K \cap L_\delta \cap G_\delta^{(1)}$ <u>for</u> $\delta \in \tilde{\gamma}$. <u>Then the following hold</u>:

(a) t_δ <u>is an involution for</u> $\delta \in \tilde{\gamma}$.

(b) $\langle t_\alpha, t_\beta\rangle$ <u>is the unique elementary abelian fours-group in</u> K.

(c) $t_\alpha = t_{\delta_3}$, $t_\beta = t_{\delta_4}$, $t_\alpha t_\beta = t_{\delta_2}$.

(d) $|C_S(t_\alpha)| = |C_S(t_\beta)|$.

<u>Proof</u>: Note that $K \cap L_\delta$, $\delta \in \tilde{\gamma}$, is a complement to some $T \in Syl_p(L_\delta)$ in $N_{L_\delta}(T)$. Hence (a) follows.

By (3.2)(c) we have $t_\delta \neq t_\lambda$ for any edge (δ,λ) in $\tilde{\gamma}$; in particular $t_\alpha \neq t_\beta$. In addition, since $K = K_\alpha K_\beta$ and K_α and K_β are cyclic and normal in K, we get (b). In particular, any involution t_δ, $\delta \in \tilde{\gamma}$, is in $\langle t_\alpha, t_\beta\rangle$.

There exists $x \in N_{L_\beta}(K)$ so that $\alpha^x = \delta_2$. Thus, the structure of $\langle t_\alpha\rangle L_\beta/Q_\beta$ implies $[t_\alpha, x] = t_\beta$ and $t_\alpha^x = t_{\delta_2} = t_\alpha t_\beta$. With the same argument $t_\beta \neq t_{\delta_3}$, and (c) follows.

Let $T \in Syl_p(L_{\delta_2} \cap L_{\delta_3})$ and set $A = C_S(t_\alpha)$ and $D = C_T(t_{\delta_3})$. Note first that $|D| = |C_S(t_\beta)|$, since β and δ_3 are conjugate. Note further that by (b) and (5.1) t_α operates fixed-point-freely on the Sylow p-subgroup of $L_\alpha \cap L_\beta/Q_\beta$ and $L_\beta \cap L_{\delta_2}/Q_{\delta_2}$ and that t_α centralizes those of $L_\alpha \cap L_\beta/Q_\alpha$ and $L_{\delta_2} \cap L_{\delta_3}/Q_{\delta_3}$ since $t_\alpha = t_{\delta_3}$. Hence

$|A| = q_\alpha q_{\delta_3} \, |C_{Q_\alpha \cap Q_{\delta_3}}(t_\alpha)|$ and with the same argument applied to t_{δ_3} $|D| = q_\alpha q_{\delta_3} \, |C_{Q_\alpha \cap Q_{\delta_3}}(t_{\delta_3})|$. Now $t_\alpha = t_{\delta_3}$ implies $|A| = |D|$ and (d) follows. □

7. The case $[Z_\alpha, Z_{\alpha'}] \neq 1$.

In this chapter we begin to determine the parameter b. This analysis will be continued into chapters 8 and 9.

(7.0) Hypothesis. In this chapter Hypothesis B' holds and furthermore $[Z_\alpha, Z_{\alpha'}] \neq 1$.

Notation. Set $\varepsilon = 1$ if $Z_\beta \neq \Omega_1(Z(S))$ and $\varepsilon = 2$ if $Z_\beta = \Omega_1(Z(S))$. Let $V = V_\alpha^{(\varepsilon)}$ and $R = [Z_\alpha, Z_{\alpha'}]$.

(7.1) Let $\delta \in \{\alpha, \alpha'\}$. Then the following hold:

(a) $L_\delta/Q_\delta \simeq SL_2(q_\delta)$, and $q_\alpha = q_{\alpha'}$.

(b) $Z_\delta/\Omega_1(Z(L_\delta))$ is a natural $SL_2(q_\delta)$-module for L_δ/Q_δ.

(c) $Z_\alpha Q_{\alpha'} \in Syl_p(L_{\alpha'})$ and $Z_{\alpha'} Q_\alpha = S$.

Proof: By (3.8)(d) and (6.2)(c) $Q_\alpha \in Syl_p(C_{L_\alpha}(Z_\alpha)) \cap Syl_p(G_\alpha^{(1)})$; in particular $Z_{\alpha'} \cap Q_\alpha = Z_{\alpha'} \cap G_\alpha^{(1)} = C_{Z_{\alpha'}}(Z_\alpha)$ and $Z_{\alpha'} \not\leq G_\alpha^{(1)}$. Hence, the hypothesis is symmetric in α and α' and $C_{Z_\alpha}(Z_{\alpha'}) = Z_\alpha \cap Q_{\alpha'}$, and it suffices to prove the claims for $\delta = \alpha$.

Set $L = \langle Z_{\alpha'}^{L_\alpha}\rangle Q_\alpha$. By (3.8)(c) and (5.12)(a) we get $|Z_{\alpha'}/Z_{\alpha'} \cap Q_\alpha| \leq |Z_\alpha/Z_\alpha \cap Q_{\alpha'}|$ and with α' in place of α

$$|Z_{\alpha'}/Z_{\alpha'} \cap Q_\alpha| = |Z_\alpha/Z_\alpha \cap Q_{\alpha'}|.$$

Now (5.12)(b) implies that $L/Q_\alpha \simeq SL_2(q_\alpha)$ and (6.1) that $L = L_\alpha$.

The remaining assertions follow again from (5.12). □

(7.2) Assume $b > 2$, then $V \not\leq G_{\alpha'}$.

Proof: Suppose that $V \leq G_{\alpha'}$. From (7.1)(c) we conclude that $V \leq Z_\alpha Q_{\alpha'}$ and hence

$$R = [Z_\alpha, Z_{\alpha'}] = [V, Z_{\alpha'}] \leq Z_\alpha .$$

Thus every subgroup of V which contains R will be normalized by $Z_{\alpha'}$. Let $\lambda \in \Delta(\alpha)$ so that $\langle Z_{\alpha'}, L_\alpha \cap L_\lambda \rangle = L_\alpha$. If $\varepsilon = 1$, then $Z_\lambda Z_\alpha$ is normal in G_α, contradicting (6.6). If $\varepsilon = 2$, then V_λ is normal in both L_λ and L_α, contradicting (3.2)(c). □

(7.3) Assume $\varepsilon = 1$, then $b \leq 2$.

Proof: Suppose to the contrary that $b > 2$. Let $T \in Syl_p(G_{\alpha'-1} \cap G_{\alpha'})$. The minimality of b implies that $V \leq L_{\alpha'-1}$ and from (7.2) we have further that $V \not\leq G_{\alpha'}$, in particular $V \not\leq Q_{\alpha'-1}$. Moreover, as $b > 2$ we have $[V,R] = 1$. Now (5.1) implies that $L_{\alpha'-1} = \langle V,T \rangle$ and R is normal in $L_{\alpha'-1}$.

Since $\varepsilon = 1$ we know by (6.2)(c) that $Q_{\alpha'-1} \in Syl_p(C_{L_{\alpha'-1}}(Z_{\alpha'-1}))$. Hence there exists $\lambda \in \Delta(\alpha)$ with $[Z_\lambda, Z_{\alpha'-1}] \neq 1$. We can now apply (7.2) to Z_λ and $Z_{\alpha'-1}$ and conclude that there is $\rho \in \Delta(\lambda)$ with $Z_\rho \leq L_{\alpha'-2}$ but $Z_\rho \not\leq Q_{\alpha'-2}$. On the other hand, as $d(\alpha,\rho) \leq 2 < b$ we have $[R, Z_\rho] = 1$. As above we have $\langle Z_\rho, P \rangle = L_{\alpha'-2}$, for $P \in Syl_p(G_{\alpha'-2} \cap G_{\alpha'-1})$, and so R

is normal in $L_{\alpha'-2}$, contradicting (3.2)(c). □

(7.4) Assume $\varepsilon = 2$, then

(a) α and α' are G-conjugate; in particular, b is even,

(b) $Z(L_\alpha) = 1$,

(c) $Z_\beta = Z_{\alpha'-1} = R$,

(d) $[V_\beta, Q_\beta] = Z_\beta$.

Proof: As $\varepsilon = 2$ we have $Z_\beta = \Omega_1(Z(L_\beta))$. But $[Z_\alpha, Z_{\alpha'}] \neq 1$ implies that $Z_{\alpha'} \nleq \Omega_1(Z(L_{\alpha'}))$ so (a) holds. From $Z_\beta \leq \Omega_1(Z(L_\beta))$ and (3.2)(c) we see that (b) holds as well. Now (7.1)(b) implies that Z_α is a natural L_α/Q_α-module and we have $R = [Z_\alpha, S] = Z_\beta$. A similar argument yields $R = Z_{\alpha'-1}$; thus (c) is proven. Finally, (d) follows as $[Q_\beta, Z_\delta] = Z_\beta$, for $\delta \in \Delta(\beta)$, and V_β is generated by conjugates of Z_δ. □

Comment: We want to conclude, in the case $\varepsilon = 2$, that $b = 2$. Looking back at (7.4)(c) we see we must deduce from $Z_{\alpha+1} = Z_{\alpha'-1}$ that $\alpha+1 = \alpha'-1$. In some cases this conclusion is easy to reach.

Suppose, for example, that G is a finite group and G_β is a maximal p-local subgroup of G. As G_β and $G_{\alpha'-1}$ are G-conjugate we have $G_\beta = N_G(Z_\beta) = N_G(Z_{\alpha'-1}) = G_{\alpha'-1}$. Now if $b > 2$ we choose $\alpha-1 \in \Delta(\alpha)$ with $G_\alpha = \langle G_{\alpha-1} \cap G_\alpha, Z_{\alpha'} \rangle$. Then we have $V_{\alpha-1} \leq Q_\beta = Q_{\alpha'-1}$ and from (7.1)(c) $[V_{\alpha-1}, Z_{\alpha'}] \leq [Z_\alpha, Z_{\alpha'}] = Z_{\alpha'-1} = Z_\beta \leq V_{\alpha-1}$. But then $V_{\alpha-1}$ is normalized by $\langle G_\alpha \cap G_{\alpha-1}, Z_{\alpha'} \rangle = G_\alpha$ and, of course, by $G_{\alpha-1}$. An application of (3.2)(c) yields a contradiction.

The following argument shows that we can, in fact, conclude that

$b = 2$ in this more general setting, although the argumentation is more involved.

(7.5) <u>Assume</u> $\varepsilon = 2$, <u>then</u> $b = 2$.

<u>Proof</u>: We assume, by way of contradiction, that $b > 2$. We immediately conclude that V_β is abelian and $[V,R] = 1$; and (7.4) implies that $b \geq 4$ and $R = Z_\beta = Z_{\alpha'-1}$. We will divide the proof into a number of shorter steps.

(1) $L_\alpha/Q_\alpha \simeq SL_2(q_\alpha)$ and Z_α is a natural L_α/Q_α-module; in particular $Z_\alpha = Z_\beta \times Z_{\alpha-1}$, $\alpha-1 \in \Delta(\alpha)\setminus\{\beta\}$, and $Z_{\alpha-1} \cap Q_{\alpha'} = 1$.

Since $\Omega_1(Z(L_\alpha)) = 1$ from (7.4)(b), the assertion follows from (7.1).

(2) $V \leq Q_{\alpha'-2}$ and $V \nleq Q_{\alpha'-1}$.

The minimality of b implies $V \leq G_{\alpha'-2}$ and by (7.4)(c) we have $Z_{\alpha'-2} = Z_{\alpha'-3}Z_{\alpha'-1} = Z_{\alpha'-2}Z_\beta$ and $[V,Z_{\alpha'-2}] = 1$. Now (1) (applied to $\alpha'-2$) yields $V \leq Q_{\alpha'-2}$, and (7.2) implies $V \nleq Q_{\alpha'-1}$.

(3) $b > 4$; in particular V operates quadratically on $V_{\alpha'-1}$.

If $b > 4$, then V is abelian and (2) implies $[V_{\alpha'-1},V,V] = [V_{\alpha'-1} \cap V,V] = 1$.

Since b is even, it remains to show that $b \neq 4$. But if $b = 4$, then by (7.4)(c) $Z_{\alpha+3} = Z_\beta$ contradicting (1) (applied to $\alpha+2$).

(4) $s \leq b-2$ (s defined in chapter 3).

Set $\tilde{\gamma} = (\alpha,\beta,\ldots,\alpha'-2)$ and let $P \in Syl_p(G_{\alpha'-2} \cap G_{\alpha'-1})$. Suppose that $b-2 \leq s-1$. From (3.4)(c) $O_p(G_{\tilde{\gamma}})$ is transitive on $\Delta(\alpha'-2) \setminus \tilde{\gamma}$ and $L_{\alpha'-2} = \langle O_p(G_{\tilde{\gamma}}),P\rangle$. On the other hand, we have $O_p(G_{\tilde{\gamma}}) \leq C_{G_\beta}(Z_\beta)$ and thus $Z_\beta = Z_{\alpha'-1}$ is normal in $G_{\alpha'-2}$, contradicting (3.2)(c).

(5) $L_{\alpha'-1}/Q_{\alpha'-1} \not\simeq L_2(q_{\alpha'-1})$ or $SL_2(q_{\alpha'-1})$.

By (3.4)(c) and (4) we may choose a path $\tilde{\gamma} = (\delta_o,\ldots,\delta_s)$ in $\gamma = (\alpha,\ldots,\alpha')$ so that $\delta_s = \alpha'-1$ or α' and $G_{\tilde{\gamma}}$ is not transitive on $\Delta(\delta_s) \setminus \{\delta_{s-1}\}$.

If $\delta_s = \alpha'$, then $Z_\alpha \leq G_{\tilde{\gamma}}$, and by (7.1)(c) Z_α operates transitively on $\Delta(\alpha') \setminus \{\alpha'-1\}$, a contradiction.

If $\delta_s = \alpha'-1$, then $V \leq G_{\tilde{\gamma}}$. In addition, by (6.7) a complement to a Sylow p-subgroup of $G_{\alpha'-2} \cap G_{\alpha'-1}$ is contained in $G_{\tilde{\gamma}}$. Thus, if $L_{\alpha'-1}/Q_{\alpha'-1} \simeq SL_2(p^n)$ or $L_2(p^n)$, then the operation of this complement on $O_p(G_{\tilde{\gamma}})$ and (5.1) show that $G_{\tilde{\gamma}}$ is transitive on $\Delta(\alpha'-1) \setminus \{\alpha'-2\}$, a contradiction.

We now derive a final contradiction. Choose $\alpha-1 \in \Delta(\alpha) \setminus \{\beta\}$ so that $V_{\alpha-1} \not\leq Q_{\alpha'-1}$ and set $|V_{\alpha-1}/V_{\alpha-1} \cap Q_{\alpha'-1}| = q$. Note that by (3) $V_{\alpha-1}$ operates quadratically on $V_{\alpha'-1}$ and that by (7.4)(d) $V_{\alpha'-1}/Z_{\alpha'-1}$

is an $L_{\alpha'-1}/Q_{\alpha'-1}$-module.

As in (7.2) we get $[V_{\alpha-1} \cap Q_{\alpha'-1}, Z_{\alpha'}] = R \leq V_{\alpha-1} \cap Q_{\alpha'-1}$. Hence, $V_{\alpha-1} \cap Q_{\alpha'-1}$ is normal in $\langle Q_{\alpha-1}, Z_{\alpha'}\rangle$, which is transitive on $\Delta(\alpha)$. Since $V_{\alpha-1} \cap V_\beta$ is also normal in $\langle Q_{\alpha-1}, Z_{\alpha'}\rangle$ we conclude that

$$V_{\alpha-1} \cap Q_{\alpha'-1} = V_{\alpha-1} \cap V_\beta = \bigcap_{\delta\in\Delta(\alpha)} V_\delta .$$

and

$$|V_{\alpha-1}/V_{\alpha-1} \cap Q_{\alpha'-1}| = |V_{\alpha-1}/V_{\alpha-1} \cap V_\beta| = q.$$

Since $(\alpha'-2, \alpha'-1)$ is conjugate to $(\alpha, \alpha-1)$, it follows that

$$|V_{\alpha'-1}/V_{\alpha'-1} \cap V_{\alpha'-3}| = q.$$

On the other hand, $[V_{\alpha-1}, V_{\alpha'-1} \cap V_{\alpha'-3}] \leq Z_{\alpha-1} \cap Z_{\alpha'-3} \leq Z_{\alpha-1} \cap Q_{\alpha'} = 1$ by (1), (2) and (3), and thus $|V_{\alpha'-1}/C_{V_{\alpha'-1}}(V_{\alpha-1})| \leq q$. Now (5.12) applied to $V_{\alpha-1}/V_{\alpha-1} \cap Q_{\alpha'-1}$ and $V_{\alpha'-1}/Z_{\alpha'-1}$ yields a contradiction to (5). □

8. The case $[Z_\alpha, Z_{\alpha'}] = 1$, Part I.

In chapter 7 we studied the case $[Z_\alpha, Z_{\alpha'}] \neq 1$. We found that this assumption holds only when $b \leq 2$. In this chapter we begin an analysis of the case $[Z_\alpha, Z_{\alpha'}] = 1$. The arguments divide quite naturally into two cases. In this first part we consider what happens when $[V_\beta \cap Q_{\alpha'}, V_{\alpha'}] \neq 1$. Our goal is to show that $b = 1$ or 3.

For the entire chapter we assume the following:

(8.0) Hypothesis. Hypothesis B' holds and furthermore:

(a) $[Z_\alpha, Z_{\alpha'}] = 1$,

(b) $[V_\beta \cap Q_{\alpha'}, V_{\alpha'}] \neq 1$,

(c) $b > 1$.

Notation. As an immediate consequence of (8.0) there exists $\lambda \in \Delta(\alpha')$ with $[V_\beta \cap Q_{\alpha'}, Z_\lambda] \neq 1$ and thus by (6.2) $V_\beta \cap Q_{\alpha'} \not\leq Q_\lambda$. Choose $T \in Syl_p(G_{\alpha'} \cap G_\lambda)$ and define $H = [V_\beta \cap Q_{\alpha'}, V_{\alpha'}]$.

(8.1) Set $\bar{L}_\lambda = L_\lambda / Q_\lambda$ and $Z = \Omega_1(Z(\bar{T}))$. Then the following hold:

(a) $Z_\beta = \Omega_1(Z(L_\beta))$, and b is odd, i.e. $b \geq 3$.

(b) $Z(L_\alpha) = 1$.

(c) $C_{L_\beta}(V_\beta) \leq G_\beta^{(1)}$ and $[V_\beta, V_{\alpha'}] \neq 1$.

(d) $Q_{\alpha'-1} \not\leq Q_{\alpha'}$.

(e) V_β operates quadratically on $V_{\alpha'}$, and vice versa; and $V_{\alpha'} \leq G_\beta$.

(f) L_δ/Q_δ is not a Ree group for every $\delta \in \Gamma$.

(g) $\langle V_\beta, T\rangle = L_{\alpha'}$.

(h) $H \leq C_{Z_\lambda}(Z)$, and Z operates quadratically on Z_λ.

(i) H is normal in $L_{\alpha'}$ and $Q_\lambda \notin \mathrm{Syl}_p(C_{L_{\alpha'}}(H))$.

Proof: From (8.0)(a) and (6.2) we have $Z_{\alpha'} = \Omega_1(Z(L_{\alpha'}))$ and from (3.8) that $Z_\alpha \nleq Z(L_\alpha)$. Hence, α and α' are not conjugate and $\alpha' \in \beta^G$; in particular (a) holds. Now $\Omega_1(Z(L_\alpha)) \leq Z_\beta$ and (3.2)(c) imply (b).

If $C_{L_\beta}(V_\beta) \nleq G_\beta^{(1)}$, then $C_{L_\beta}(V_\beta)$ and thus also $C_{L_\beta}(Z_\alpha)$ is transitive on $\Delta(\beta)$, contradicting (3.2)(c). Since $\alpha' \in \beta^G$ we also have $C_{L_{\alpha'}}(V_{\alpha'}) \leq G_{\alpha'}^{(1)}$ and $[V_\beta, V_{\alpha'}] \neq 1$ since $V_\beta \nleq G_{\alpha'}^{(1)}$.

The minimality of b yields $V_{\alpha'} \leq L_\beta$ and $V_\beta \leq Q_{\alpha'-1}$. The second inclusion gives (d), the first inclusion together with $b \geq 3$ gives (e). With the same argument $V_\beta \cap Q_{\alpha'}$ operates quadratically on Z_λ. Now (5.9) implies (f).

Assertion (g) is a consequence of (5.1).

By (5.9) $\overline{V_\beta \cap Q_{\alpha'}} \leq Z$. Set $H_0 = [V_\beta \cap Q_{\alpha'}, Z_\lambda]$. Then H_0 is normalized by T. As H_0 also lies in V_β, we get from (g) that H_0 is normal in $L_{\alpha'}$; in particular $H_0 \leq \bigcap_{\delta\in\Delta(\alpha')} Z_\delta = X$ and X is normal in $L_{\alpha'}$. Hence $H \leq X \leq Z_\lambda$. Now H is normalized by T and by V_β. By (g) H is normal in $L_{\alpha'}$.

Set $W = \langle Q_\lambda^{L_{\alpha'}}\rangle$, then $H \leq Z(W)$ and by (6.5) $Q_\lambda \notin \mathrm{Syl}_p(C_{L_{\alpha'}}(H))$. Hence (i) holds.

Finally, let $\tilde{K}$ be a complement to T in $G_{\alpha'} \cap G_\lambda$. Then $\tilde{K}$ normalizes W and $Z_\lambda \cap Z(W)$, and the operation of $\tilde{K}$ on $\bar{T}$ gives

$[\overline{V_\beta \cap Q_{\alpha'}},\tilde{K}] = Z$ or $|Z| = |V_\beta \cap Q_{\alpha'}| = 2$. Since $H_o \leq Z_\lambda \cap Z(W)$ and $V_\beta \cap Q_{\alpha'} \leq W$ this shows that $[Z_\lambda, Z] \leq Z(W)$ and $[Z_\lambda, Z, Z] = 1$, and (h) follows. □

Remark: We will use (8.1)(a) - (g) without reference and sometimes "after conjugation", i.e. $Z_\delta = \Omega_1(Z(L_\delta))$ for $\delta \in \beta^G$ (in particular for $\delta = \alpha'$), $Z(L_\delta) = 1$ for $\delta \in \alpha^G$ (in particular for $\delta = \lambda$), etc.

(8.2) $V_{\alpha'} \not\leq Q_\beta$.

Proof: Assume, by way of contradiction, that $V_{\alpha'} \leq Q_\beta$. Let $S \in Syl_p(G_\alpha \cap G_\beta)$, $\bar{L}_\alpha = L_\alpha/Q_\alpha$ and $Z = \Omega_1(Z(\bar{S}))$. Note that by (8.1)(e) Z operates quadratically on Z_α and by (5.9) $\bar{V}_{\alpha'} \leq Z$.

Suppose first that $|Z_\alpha/C_{Z_\alpha}(Z)| < |Z|^2$. From (8.1)(b) and (5.12) we have that $\bar{L}_\alpha \simeq SL_2(|Z|)$ and Z_α is a natural $\bar{L}_\alpha$-module. Then Z_λ is also a natural L_λ/Q_λ-module. Hence, $H = Z_{\alpha'} \leq [V_{\alpha'}, V_\beta] = Z_\beta$. We conclude that $Z_{\alpha'} = Z_\beta = [V_{\alpha'}, V_\beta]$. But now $[\langle V_\beta^{L_{\alpha'}}\rangle, V_{\alpha'}] \leq Z_{\alpha'}$ and Z_λ is normal in $\langle V_\beta^{L_{\alpha'}}\rangle$, contradicting (3.2)(c).

Thus, we may assume now that

$$(*) \qquad |Z_\alpha/C_{Z_\alpha}(Z)| \geq |Z|^2.$$

According to (5.10) we then also have

$$(**) \qquad |Z_\alpha/C_{Z_\alpha}(A)| \geq |Z|^2 \quad \text{for} \quad 1 \neq A \leq Z.$$

If $[Z_\alpha \cap Q_{\alpha'}, V_{\alpha'}] = 1$, then $Z_\alpha/C_{Z_\alpha}(V_{\alpha'}) = Z_\alpha/Z_\alpha \cap Q_{\alpha'}$ and by (5.12)(a) applied to $L_{\alpha'}$ and a suitable factor module of $V_{\alpha'}$ we get $|Z_\alpha/Z_\alpha \cap Q_{\alpha'}| \leq |V_{\alpha'}/V_{\alpha'} \cap Q_\alpha| = |\bar{V}_{\alpha'}|$. But now (**) is violated since by (5.9) $\bar{V}_{\alpha'} \leq Z$.

So suppose that $[Z_\alpha \cap Q_{\alpha'}, V_{\alpha'}] \neq 1$. We may assume that $Z_\alpha \cap Q_{\alpha'} \not\leq Q_\lambda$. Since λ is G-conjugate to α, we get from (**) that $|\bar{Z}_\lambda| \geq |Z|^2$. But this absurd since $\bar{Z}_\lambda \leq Z$. □

(8.3) <u>Let</u> $Z = \Omega_1(Z(T/Q_\lambda))$, <u>then</u> $[Z, Z_\lambda] \leq H$.

<u>Proof</u>: Let L_λ be a counter example and set $\bar{L}_\lambda = L_\lambda/Q_\lambda$. Obviously then $Z \neq \overline{V_\beta \cap Q_{\alpha'}}$ and thus $q_\lambda > 2$. By (8.1)(a) $Z(L_\lambda) = 1$, and (8.1)(h), (5.2), (5.3) and (5.8)(d) imply:

(1) $\bar{L}_\lambda \simeq U_3(q_\lambda)$ or $SU_3(q_\lambda)$.

Let $\tilde{K}$ be a complement to T in $G_{\alpha'} \cap G_\lambda$. Furthermore, we define:

$$L = O^p(L_{\alpha'}), \quad T_o = T \cap LQ_\lambda, \quad \tilde{K}_o = \tilde{K} \cap L,$$
$$V = V_{\alpha'}/C_{V_{\alpha'}}(L), \quad A = C_{Z_\lambda}(L),$$
$$a_1 = |V_\beta/V_\beta \cap Q_{\alpha'}|, \quad b_1 = |V_\beta \cap Q_{\alpha'}/V_\beta \cap Q_\lambda|,$$
$$a_2 = |V_{\alpha'}/V_{\alpha'} \cap Q_\beta|, \quad b_2 = |V_{\alpha'} \cap Q_\beta/V_{\alpha'} \cap Q_\alpha|.$$

Note that by (8.1) and the operation of $\tilde{K}$ on Z we have $[Z_\lambda, Z] \leq A \leq C_{Z_\lambda}(Z)$ and $Z \leq \bar{T}_o$. Note further that $\tilde{K}_o$ is a complement to T_o in $G_\lambda \cap LQ_\lambda$.

Assume first that $\bar{T}_o \neq Z$. Then the operation of $\tilde{K}$ on T_o yields $T_o = T$. Let $g \in L_\lambda \setminus G_{\alpha'}$ and $Y = \langle Z, Z^g \rangle$. By (5.8)(d) we get $C_{Z_\lambda}(Z) = RH$ where $R = C_{Z_\lambda}(Y)$. Since $A \neq H$ it follows that $R \cap A \neq 1$. On the other hand, $\langle \bar{T}_o, Z^g \rangle = \bar{L}_\lambda$ centralizes $A \cap R$ which contradicts $Z(L_\lambda) = 1$.

Thus, we have $\bar{T}_o = Z$. In particular, $[Z_\lambda, O_p(LT_o)] \leq A$ and V is an $LT_o/O_p(LT_o)$-module.

Assume that $|V/C_V(V_\beta)| < a_1^2$. Then by (5.12) $L_{\alpha'}/Q_{\alpha'} \simeq SL_2(q_{\alpha'})$ and V is a natural $SL_2(q_{\alpha'})$-module. Note that $\tilde{K}$ operates irreducibly on Z_λ/A and Z. It follows that $A = C_{Z_\lambda}(Z)$ and $|Z_\lambda/A| = q_{\alpha'}$. In addition, (8.1)(h), (1) and (5.12) yield $|Z|^2 \leq q_{\alpha'}$. Since $\tilde{K}_o$ is cyclic of order $q_{\alpha'}-1$ ($\neq 1$) and $\tilde{K}$ normalizes $\tilde{K}_o$ there exists $1 \neq k \in \tilde{K}_o$ so that $[Z,k] = 1$. But now the Three-subgroup-Lemma implies $[k, Z_\lambda] \leq A$, contradicting the operation of k on V.

We have shown that $a_1^2 \leq |V/C_V(V_\beta)|$. If $C_V(V_\beta) = C_V(Z_\alpha)$, then clearly $|V/C_V(V_\beta)| \leq a_2 b_2$; and if $C_V(V_\beta) \neq C_V(Z_\alpha)$, then (5.10) yields $a_2 b_2 \geq |V/C_V(Z_\alpha)| \geq q_{\alpha'}^2$. Note that by (8.2) the same argument can be applied to $V_\beta/C_{V_\beta}(O^p(L_\beta))$ and $V_{\alpha'}$. Since $a_1 \leq q_{\alpha'}$ and $b_2 \leq q_\lambda = q_\alpha$ we get:

(2) $a_1^2 \leq a_2 b_2 \leq a_2 q_\lambda$ and $a_2^2 \leq a_1 q_\lambda$.

From (1) and (5.10) we get:

(3) $q_\lambda^2 \leq |Z_\lambda / C_{Z_\lambda}(V_\beta \cap Q_{\alpha'})|$.

Assume that $[V_\beta, Z_\lambda \cap Q_\beta] = 1$. Then (3) implies $q_\lambda^2 \leq a_2$. But now we conclude from (2) that $a_1^2 \leq a_2 q_\lambda < a_2^2$ and $a_2^2 \leq a_1 q_\lambda < a_1 a_2$ which is absurd.

Assume now that $[Z_\lambda \cap Q_\beta, V_\beta] \neq 1$. Then there exists $\tilde{\alpha} \in \Delta(\beta)$ with $[Z_{\tilde{\alpha}}, Z_\lambda \cap Q_\beta] \neq 1$. Now (1) and (5.10) applied to $L_{\tilde{\alpha}}$ yield

(4) $q_\lambda^2 = q_{\tilde{\alpha}}^2 \leq a_1 q_\lambda$.

It follows that $q_\lambda \leq a_1$ and together with (2) that $q_\lambda \leq a_1 \leq a_2 \leq q_\lambda$. This shows that $a_2 = q_\lambda$ and $\overline{V_\beta \cap Q_{\alpha'}} = Z$, and L_λ is not a counter example. □

(8.4) $b = 3$.

<u>Proof:</u> Assume that $b > 3$ and set $W = V_\beta^{(3)}$. Then $[W,H] = 1$ and the minimality of b implies $W \leq L_{\alpha'-2}$.

(1) $W \leq Q_{\alpha'-1}$.

From (8.1)(h) and (i) we have that $H \leq Z_{\alpha'-1}$ and so H is normalized by $\langle W \cap L_{\alpha'-1}, L_{\alpha'-1} \cap L_{\alpha'}\rangle$. Now (3.2)(c) yields $W \cap L_{\alpha'-1} \leq Q_{\alpha'-1}$. Thus, we may suppose that $W \not\leq Q_{\alpha'-2}$. Now H is centralized by $\langle W, Q_{\alpha'-1}\rangle = X$. In particular, $L_{\alpha'-2}/Q_{\alpha'-2} \simeq U_3(q_{\alpha'-2})$ or $SU_3(q_{\alpha'-2})$ and $L_{\alpha'-2}$ is doubly transitive on $\Delta(\alpha'-2)$. In addition, as above, $X \cap L_{\alpha'-1} \leq G_{\alpha'}$ and so $Q_{\alpha'-1} \in Syl_p(X)$. It follows that $Q_{\alpha'-1} \cap Q_{\alpha'-2} \leq Q_{\alpha'-3}$. Hence, after conjugation to α' and using the 2-fold transitivity of $L_{\alpha'}$ on $\Delta(\alpha')$ we get $Q_{\alpha'-1} \cap Q_{\alpha'} \leq Q_\lambda$. But this contradicts $V_\beta \cap Q_{\alpha'} \not\leq Q_\lambda$.

(2) If $Q_\alpha Q_\beta = S$ and $L_\delta/Q_\delta \not\cong D_{10}$ for $\delta \in \Gamma$, then W is elementarytary abelian.

If $b > 5$, then $b \geq 7$ by (8.1)(a) and the generators of W centralize each other giving (2) in this instance. So let $b = 5$. We have $Z_{\alpha'-1} \leq W$ and from (1) $Z_{\alpha'-1} \leq Z(W)$. On the other hand, by assumption and (8.1)(f), $L_\delta/Q_\delta \not\cong D_{10}$ or $Ree(3)'$. It follows that a Sylow p-subgroup of $L_\delta \cap L_\mu$ $(\mu \in \Delta(\delta))$ is already transitive on $\Delta(\delta) \setminus \{\mu\}$. Since $S = Q_\alpha Q_\beta$ we conclude that L_β is transitive on the paths of length 3 beginning at β, and (2) follows.

Let A and B be the preimages of $\Omega_1(Z(\tilde{T}/Q_{\alpha'}))$, $\tilde{T} \in Syl_p(L_{\alpha'} \cap L_{\alpha'-1})$, and $\Omega_1(Z(T/Q_\lambda))$, respectively.

(3) $W \leq A$ and $W \cap Q_{\alpha'} \leq B$.

If $p = 2$ or $Q_{\alpha'} Q_{\alpha'-1} \notin Syl_p(L_{\alpha'})$, then (3) follows. Hence, we can apply (2) and get $\phi(W) = 1$. In particular, W operates quadratically on $V_{\alpha'}$, and the assertion follows from (5.9)(b).

We now complete the proof as follows: By (8.3), (3) and (8.1)(i) $[W \cap Q_{\alpha'}, V_{\alpha'}] \leq H \leq V_\beta$. Hence, $(W \cap Q_{\alpha'})V_\beta/V_\beta$ is centralized by $V_{\alpha'}$, and again by (3) $|W/(W \cap Q_{\alpha'})V_\beta| < q_{\alpha'} = q_\beta$. Now (8.2) and (5.7) imply $[W, O^p(L_\beta)] \leq V_\beta$. However, we can now conclude that $V_\alpha^{(2)}$ is normal in both L_α and $O^p(L_\beta)$ which contradicts (3.2)(c). □

(8.5) Let $Z = C_{V_\beta}(O^p(L_\beta))$ and $L_\beta/Q_\beta \simeq SL_2(q_\beta)$. Then V_β/Z is not a natural L_β/Q_β-module.

Proof: By (8.4) we have $\alpha'-1 = \alpha+2$, and by (8.1) H is normal in $L_{\alpha'}$ and $Z_{\alpha'} \leq Z(L_{\alpha'})$. It follows that $H \cap Z_{\alpha'} \neq 1$. Let $\tilde{K} = L_\beta \cap G_{\alpha+2} \cap G_{\alpha'}$. If V_β/Z is a natural L_β/Q_β-module, then $\tilde{K}$ is irreducible on $V_{\alpha'}/V_{\alpha'} \cap Q_\beta$, $V_\beta/V_\beta \cap Q_{\alpha'}$ and $V_\beta \cap Q_{\alpha'}/Z$. Hence, $H = [V_\beta \cap Q_{\alpha'}, V_{\alpha'}] \leq Z$ and $H \cap Z_{\alpha'}$ is centralized by $\langle O^p(L_\beta) \cap G_{\alpha+2}, G_{\alpha+2} \cap L_{\alpha'}\rangle$. Now (3.2)(c) implies that $Q_{\alpha+2} \in Syl_p(Q_{\alpha+2}O^p(L_\beta))$, contradicting (6.5). □

(8.6) The following statements are equivalent:

(a) $H \leq Z_{\alpha'}$.

(b) $H \cap H^g = 1$ for $g \in G_{\alpha+2} \setminus G_{\alpha'}$.

(c) $[V_\beta \cap Q_{\alpha'}, V_{\alpha'} \cap Q_\beta] = 1$.

(d) $L_\alpha/Q_\alpha \simeq SL_2(q_\alpha)$, Z_α is a natural L_α/Q_α-module, and $H = Z_{\alpha'}$.

Proof: By (8.1) and (8.4) we have $\alpha'-1 = \alpha+2$, $Z_{\alpha'} = \Omega_1(Z(L_{\alpha'}))$ and $Z(L_{\alpha+2}) = 1$. Hence, (a) implies (b) by the structure of $L_{\alpha+2}/Q_{\alpha+2}$. From (b) we get (c) since $[V_\beta \cap Q_{\alpha'}, V_{\alpha'} \cap Q_\beta] \leq H \cap H^g$ where $g \in G_{\alpha+2}$ with $\alpha'^g = \beta$.

If $L_\alpha/Q_\alpha \simeq SL_2(q_\alpha)$ and Z_α is a natural L_α/Q_α-module, then L_λ has the same property and $H = Z_{\alpha'}$, and (a) holds.

Assume now that (c) holds but not (d). Let $\bar{L}_\lambda = L_\lambda/Q_\lambda$ and $Z = \Omega_1(Z(\bar{T}))$. By (8.1)(h) and (5.12) we have $|Z_\lambda/C_{Z_\lambda}(Z)| \geq q_\lambda^2$. Now (5.10) applied to $\overline{V_\beta \cap Q_{\alpha'}} \leq Z$ yields either $C_{Z_\lambda}(Z) = C_{Z_\lambda}(V_\beta \cap Q_{\alpha'})$ or $|Z_\lambda/C_{Z_\lambda}(V_\beta \cap Q_{\alpha'})| \geq q_\lambda^2$. In both cases we get:

$$q_\lambda^2 \leq |Z_\lambda/C_{Z_\lambda}(V_\beta \cap Q_{\alpha'})| \leq |Z_\lambda/Z_\lambda \cap Q_\beta| \leq q_\beta.$$

On the other hand, since $b = 3$ the operation of K (or a suitable $L_{\alpha+2}$-conjugate) shows that $|V_\beta/V_\beta \cap Q_{\alpha'}| = q_{\alpha'}$, and we deduce from (8.5) and (5.10) that $q_{\alpha'}^2 \leq q_\alpha q_\beta$. Since $q_{\alpha'} = q_\beta$ and $q_\lambda = q_\alpha$ we conclude that $q_\lambda < q_\beta \leq q_\lambda$, a contradiction. □

(8.7) $H = Z_{\alpha'}$ <u>and</u> $q_\alpha = q_\beta$.

<u>Proof</u>: Assume first that $H \leq Z_{\alpha'}$. Then all the statements in (8.6) hold, and we get as in the proof of (8.6): $q_\lambda \leq q_\beta$, $q_{\alpha'}^2 \leq q_\alpha q_\beta$, and thus $q_\alpha = q_\beta$. Hence, it suffices to show $H \leq Z_{\alpha'}$.

We may assume that none of the statements in (8.6) holds. As in the proof of (8.6) we have: $q_{\alpha'}^2 \leq q_\alpha q_\beta$, $q_{\alpha'} \leq q_\alpha$, and $q_\lambda^2 \leq |Z_\lambda/C_{Z_\lambda}(V_\beta \cap Q_{\alpha'})|$. It follows that $C_{Z_\lambda}(V_\beta \cap Q_{\alpha'}) \nleq Q_\beta$, and by (5.10) and (8.5) applied to L_β we get $q_\beta^2 \leq |V_\beta/V_\beta \cap Q_{\alpha'}| \leq q_{\alpha'}$, a contradiction. □

9. The case $[Z_\alpha, Z_{\alpha'}] = 1$, Part II.

In this chapter we complete the analysis of the case $[Z_\alpha, Z_{\alpha'}] = 1$. For this purpose we assume for the entire chapter that the following hypothesis holds.

(9.0) Hypothesis. Hypothesis B' holds and furthermore:

(a) $[Z_\alpha, Z_{\alpha'}] = 1$,

(b) $[V_\beta \cap Q_{\alpha'}, V_{\alpha'}] = 1$,

(c) $b > 1$.

Our goal is to show that under (9.0) $b = 3$ or 5. In a later chapter this will lead to the conclusion that G is parabolic isomorphic to M_{12} (for $b = 3$), or of type F_3 (for $b = 5$). Thus no "classical" families arise under these conditions. Propositions (9.10) and (9.11) where it is shown that $q_\alpha = 2$, resp. $q_\alpha = 3$, when $b = 3$, resp. $b = 5$, may give some explanation for the exceptional nature of the structure of G_α and G_β.

As in (8.1) it is immediate that the following holds:

(9.1)(a) $b \equiv 1(2)$, i.e. α' is conjugate to β and $q_\beta = q_{\alpha'}$.

(b) $Z(L_\alpha) = 1$ and $Z_{\alpha'} = \Omega_1(Z(L_{\alpha'}))$.

(c) V_β is elementary abelian and operates quadratically on $V_{\alpha'}$, and vice versa.

(d) $C_{L_\beta}(V_\beta) \leq G_\beta^{(1)}$. □

We will use (9.1) without reference.

(9.2) <u>Suppose that</u> $V_{\alpha'} \not\leq Q_\beta$, <u>then</u> $[V_{\alpha'} \cap Q_\beta, V_\beta] = 1$.

<u>Proof</u>: Assume to the contrary that $[V_{\alpha'} \cap Q_\beta, V_\beta] \neq 1$ and choose $\lambda \in \Delta(\alpha')$ such that $Z_\lambda \not\leq Q_\beta$. Then λ and β in place of α and α' satisfy (8.0). It follows from (8.4), (8.6), and (8.7) that $b = 3$ and that $L_{\alpha+2}/Q_{\alpha+2} \simeq SL_2(q_{\alpha+2})$. Thus α' and β are in $\Delta(\alpha+2)$ and there exists $t \in L_{\alpha+2}$ with $\beta^t = \alpha'$ and $(\alpha')^t = \beta$. This yields $[V_{\alpha'} \cap Q_\beta, V_\beta]^t = [V_\beta \cap Q_{\alpha'}, V_{\alpha'}] \neq 1$, a contradiction. □

(9.3)(a) $L_\beta/Q_\beta \simeq SL_2(q_\beta)$,

(b) $|V_\beta/V_\beta \cap Q_{\alpha'}| = q_{\alpha'}$ <u>and</u> $|V_{\alpha'}/C_{V_{\alpha'}}(V_\beta)| = q_{\alpha'}$.

(c) $V_\beta/C_{V_\beta}(O^p(L_\beta))$ <u>is a natural</u> L_β/Q_β<u>-module</u>.

<u>Proof</u>: Assume first that $V_{\alpha'} \not\leq Q_\beta$. Note that $C_{V_\beta}(V_{\alpha'}) = V_\beta \cap Q_{\alpha'}$ and by (9.2) $C_{V_{\alpha'}}(V_\beta) = V_{\alpha'} \cap Q_\beta$. Now (5.12) (applied to L_β and $L_{\alpha'}$) implies the assertions.

Assume next that $V_{\alpha'} \leq Q_\beta$. Clearly, $V_{\alpha'} \not\leq Q_\alpha$ since $Z_\alpha \not\leq G_{\alpha'}^{(1)}$. So we have $C_{Z_\alpha}(V_{\alpha'}) = Z_\alpha \cap Q_{\alpha'}$ and $C_{V_{\alpha'}}(Z_\alpha) = V_{\alpha'} \cap Q_\alpha$ and again (5.12) (applied to L_α and $L_{\alpha'}$) yields (a) and (c) after conjugating α' to β, and $|Z_\alpha/Z_\alpha \cap Q_{\alpha'}| = q_{\alpha'} = |V_{\alpha'}/V_{\alpha'} \cap Q_\alpha|$. It follows that $V_\beta = Z_\alpha(V_\beta \cap Q_{\alpha'})$ and (b) holds, too. □

(9.4)(a) $L_\alpha/Q_\alpha \simeq SL_2(q_\alpha)$ <u>and</u> Z_α <u>is a natural</u> L_α/Q_α<u>-module</u>.

(b) $q_\alpha = q_\beta$.

(c) $[Q_\beta, V_\beta] = Z_\beta$.

<u>Proof</u>: Set $W = \langle Q_\alpha^{G_\beta}\rangle$, $Z = Z_\alpha \cap Z(W)$ and $A = \Omega_1(Z(S/Q_\alpha))$. Then $WQ_\beta = L_\beta$ and $K_\beta \leq W$. Note that $Z = C_{Z_\alpha}(O^p(L_\beta))$ and by (6.5) $W \cap S \nleq Q_\alpha$. Hence, we get from (9.3) and the operation of K on W and Z_α:

(i) $|Z_\alpha/Z| = q_\beta$.

(ii) $Z = C_{Z_\alpha}(k)$ for $1 \neq k \in K_\beta$.

(iii) The preimage of A is in W.

In particular, (ii) and (iii) imply that A operates quadratically on Z_α and by (5.12) and (i) $q_\alpha \leq q_\beta$.

If $q_\alpha = q_\beta$, then again (5.12) yields (a) and (b), and (c) is an easy consequence of (a). Thus, we may assume that $q_\alpha < q_\beta$.

There exists a conjugate $A^g \neq A$ in L_α/Q_α which is normalized by K_β. For $L = \langle A, A^g\rangle$ we get from (5.2) - (5.4) that A and A^g are conjugate in L, $O_p(L) = 1$, and $L = \langle x, A\rangle$ for $1 \neq x \in A^g$. Since $|K_\beta| = q_\beta - 1$ and $q_\alpha < q_\beta$ there are $1 \neq k \in K_\beta$ and $1 \neq x \in A^g$ so that $[x,k] = 1$. By the Three-subgroup Lemma we get $[Z,x,k] = 1$ and by (ii) $[Z,x] \leq Z$. Hence, L centralizes Z_α/Z and Z. This implies that $O^p(L) \leq C_{L_\alpha/Q_\alpha}(Z_\alpha) = 1$, which contradicts $O_p(L) = 1$. □

In the following lemmata we use the parameter s which was defined in chapter 3.

(9.5) <u>Suppose that</u> $s < b$ <u>and</u> $\tilde\gamma = (\alpha_0, \ldots, \alpha_s)$ <u>is a path of length</u> s <u>with</u> $O_p(G_{\tilde\gamma}) \nleq G_{\alpha_s}^{(1)}$. <u>Then</u> α_s <u>is conjugate to</u> α'.

<u>Proof</u>: By the definition of s there exists a path $\hat{\gamma} = (\hat{\alpha}_o,\ldots,\hat{\alpha}_s)$ of length s so that $G_{\hat{\gamma}}$ is not transitive on $\Delta(\hat{\alpha}_s)\setminus\{\hat{\alpha}_{s-1}\}$. By (3.4)(c) and since $b > s$ we may assume (after conjugation) that

$$\hat{\gamma} \subseteq \gamma \quad \text{and} \quad \hat{\alpha}_s \in \{\alpha'-1,\alpha'\}.$$

It follows from (6.8) together with (9.3)(a) and (9.4)(a) that $O_p(G_{\hat{\gamma}}) \leq G^{(1)}_{\hat{\alpha}_s}$. Hence, $\tilde{\gamma}$ is not conjugate to $\hat{\gamma}$ and thus α_s is not conjugate to $\hat{\alpha}_s$; i.e. α_s is in the same orbit as $\{\alpha',\alpha'-1\}\setminus\{\hat{\alpha}_s\}$.

If $\hat{\alpha}_s = \alpha'$, then $V_\beta \leq G_\gamma \leq G_{\hat{\gamma}}$ but $V_\beta \not\leq G^{(1)}_{\alpha'}$, a contradiction. Hence $\hat{\alpha}_s = \alpha'-1$ and $\alpha_s \in \alpha'^G$. □

(9.6) $V_{\alpha'} \not\leq Q_\beta$ <u>and</u> $|V_{\alpha'}/V_{\alpha'} \cap Q_\beta| = q_\beta$.

<u>Proof</u>: If $s \geq b$, then by (3.4)(c) $(\alpha',\ldots,\beta)$ is conjugate to $(\beta,\ldots,\alpha')$ and the assertion follows.

Suppose now that $s < b$ and $V_{\alpha'} \leq Q_\beta$. By (9.5) $V_{\alpha'} \leq G^{(1)}_\alpha$, since α is not conjugate α', and $[V_{\alpha'},Z_\alpha] = 1$ contradicting (9.1)(d).

Thus, we have $V_{\alpha'} \not\leq Q_\beta$ and according to (9.2) the pair (α',β) has the same properties as (β,α'). Hence, $|V_{\alpha'}/V_{\alpha'} \cap Q_\beta| = q_\beta$ follows from (9.3)(b) by reversing β and α'. □

(9.7) <u>Set</u> $W = V^{(2)}_\alpha$. <u>Then</u> $W \leq L_{\alpha'-2}$ <u>but</u> $W \not\leq Q_{\alpha'-2}$.

<u>Proof</u>: Obviously, the minimality of b implies $W \leq L_{\alpha'-2}$. Pick $\alpha-1 \in \Delta(\alpha)\setminus\{\beta\}$ and $\alpha-2 \in \Delta(\alpha-1)\setminus\{\alpha\}$. If $s \geq b$, then the path

$(\alpha-2,\alpha-1,\alpha,\ldots,\alpha'-2)$ is conjugate to γ and thus $V_{\alpha-1} \not\leq Q_{\alpha'-2}$.

Hence, we may assume that $s < b$ and $W \leq Q_{\alpha'-2}$. Note that $\tilde{\gamma} = (\alpha-1,\alpha,\ldots,\alpha'-1)$ contains a subpath of length s. It follows by (9.5) that $W \leq G_{\tilde{\gamma}} \leq G^{(1)}_{\alpha'-1}$, and (9.3) yields $W \leq V_\beta Q_{\alpha'}$ and $W = V_\beta(W \cap Q_{\alpha'})$. Again by (9.5) we get $W \cap Q_{\alpha'} \leq G^{(1)}_\delta$ for every $\delta \in \Delta(\alpha')$ and thus $[W \cap Q_{\alpha'},V_{\alpha'}] = 1$. We conclude that $[W,V_{\alpha'}] = [V_\beta,V_{\alpha'}] \leq V_\beta \leq W$. It follows together with (9.6) that W is normal in L_α and L_β contradicting (3.2)(c). □

(9.8) $[V_\beta,V_{\alpha'}] \leq Z_{\alpha'-1} \cap Z_{\alpha+2}$, $V_\beta = Z_\alpha Z_{\alpha+2}$ <u>and</u> $|V_\beta| = q_\alpha^3$.

<u>Proof</u>: Set $R = [V_\beta,V_{\alpha'}]$ and $\bar{V}_\beta = V_\beta/Z_\beta$. By (9.4)(c) $\bar{V}_\beta$ is an L_β/Q_β-module, and by (9.3) $\bar{V}_\beta/C_{\bar{V}_\beta}(L_\beta)$ is a natural $SL_2(q_\beta)$-module. Thus, we have $V_\beta = RR^gZ_{\alpha+2}$ for $g \in L_\beta$ with $(\alpha+2)^g = \alpha$.

If $V_\beta = Z_\alpha Z_{\alpha+2}$, then $|V_\beta| = q_\alpha^3$ and $R \leq Z_{\alpha+2}$, and since α' is conjugate to β, we also have $R \leq Z_{\alpha'-1}$. Hence, it suffices to show $V_\beta = Z_\alpha Z_{\alpha+2}$ or $R \leq Z_{\alpha+2} \cap Z_{\alpha'-1}$.

Suppose first that $b = 3$. Then by (9.4)(c), (9.3)(b) and (9.6) $RZ_{\alpha+2}$ is normal in $\langle Q_\beta,Q_{\alpha'}\rangle Q_{\alpha+2} = L_{\alpha+2}$ and $[RZ_{\alpha+2},L_{\alpha+2}] \leq Z_{\alpha+2}$. On the other hand, $Q_{\alpha+2} = V_\beta(Q_{\alpha+2} \cap Q_{\alpha'})$ and $[RZ_{\alpha+2},Q_{\alpha+2}] = [R,Q_{\alpha+2} \cap Q_{\alpha'}] \leq Z_{\alpha'}$. Furthermore, $[RZ_{\alpha+2},Q_{\alpha+2}]$ is normal in $L_{\alpha+2}$. It follows together with (3.2)(c) that $RZ_{\alpha+2} \leq Z(Q_{\alpha+2})$.

Note that by (9.4)(b) $q_\alpha = q_\beta$. If $q_\beta = 2$, then $L_{\alpha+2}/Q_{\alpha+2} \simeq SL_2(2)$ and $RZ_{\alpha+2} \cap Z(L_{\alpha+2}) \neq 1$, a contradiction. Thus, we have $q_\beta > 2$ and therefore $K_\beta \neq 1$. By (6.8)(b) we may choose K_β to be in $G_{\alpha+2} \cap G_{\alpha'}$. Hence, K_β normalizes R and $Z_{\alpha+2}$.

Assume that $[K_\beta,L_{\alpha+2}] \leq Q_{\alpha+2}$. Then $L_{\alpha+2}$ normalizes $[Z_{\alpha+2}R,K_\beta]$.

Since $Z_{\alpha+2} \not\leq [Z_{\alpha+2}R,K_\beta]$ we conclude from (9.4)(a) that $[Z_{\alpha+2}R,K_\beta] \leq Z(L_{\alpha+2}) = 1$. But this contradicts (9.3)(c) and the operation of $SL_2(q_\beta)$ on a natural module.

Assume now that $[K_\beta,L_{\alpha+2}] \not\leq Q_{\alpha+2}$. By (9.4)(a) and (5.1)(g) we have $S^{g^{-1}} = [Q_\beta,K_\beta]Q_{\alpha+2}$, g as above. Set $R_o = C_R(O^p(L_\beta))$. Then R_o is K_β-invariant and $C_{L_\beta}(R_o)$ operates transitively on $\Delta(\beta)$. Hence, it follows from (6.5) that $R_o \leq \Omega_1(Z(S^{g^{-1}})) = Z_\beta$. In particular, $\bar{R} = [\bar{R},K_\beta]$ and $|\bar{R}| = q_\beta$, and by (9.3)(c) $\bar{R}$ is the only non-trivial K_β-submodule in $\overline{V_\beta \cap Q_{\alpha'}}$. Hence, we either have $\bar{R} = \bar{Z}_{\alpha+2}$ or $[\bar{Z}_{\alpha+2},K_\beta] = 1$. In the first case the assertion follows; in the second case we get $[Z_{\alpha+2},K_\beta] = 1$ and thus $[L_{\alpha+2},K_\beta] \leq Q_{\alpha+2}$, a contradiction.

Suppose now that $b > 3$ and define W as in (9.7). Since $R \leq V_\beta$ we have $[R,W] = 1$. By (9.7) there exists a path $(\alpha-2,\alpha-1,\alpha)$ so that $Z_{\alpha-2} \not\leq Q_{\alpha'-2}$. In view of (8.4) the pair $(\alpha-2,\alpha'-2)$ and $\alpha-1$ fulfil Hypothesis (9.0) in place of (α,α') and β. Hence, by (9.3) we have $V_{\alpha-1}Q_{\alpha'-2} = WQ_{\alpha'-2} \in Syl_p(L_{\alpha'-2})$.

Since $s \geq 2$ there exists $g \in G_{\alpha'-2}$ so that $(\alpha'-2,\alpha'-1,\alpha')$ is on the path $(\alpha'-2,\ldots,\alpha^g)$ and therefore $[V_{\alpha'},W^g] = 1$. Note that $\langle W,W^g\rangle Q_{\alpha'-2} = L_{\alpha'-2}$ and by (6.5) $Q_{\alpha'-2} = (Q_{\alpha'-2} \cap Q_{\alpha'-3})(Q_{\alpha'-2} \cap Q_{\alpha'-1})$. It follows that $L_{\alpha'-2} = C_{L_{\alpha'-2}}(R)Q_{\alpha'-2}$. In addition, as $Q_{\alpha'-2} \cap Q_{\alpha'-1} = V_\beta(Q_{\alpha'-2} \cap Q_{\alpha'})$ we have $[R,Q_{\alpha'-2} \cap Q_{\alpha'-1}] \leq Z_{\alpha'} \leq Z_{\alpha'-1}$ and thus $[R,Q_{\alpha'-2} \cap Q_{\alpha'-3}] \leq Z_{\alpha'-3}$ and $[R,Q_{\alpha'-2}] \leq Z_{\alpha'-1}Z_{\alpha'-3}$. Since $[R,Q_{\alpha'-2}]$ is a normal subgroup of $L_{\alpha'-2}$ we get by (9.3)(c) and (9.4)(b) either $[R,Q_{\alpha'-2}] = Z_{\alpha'-1}Z_{\alpha'-3} = V_{\alpha'-2}$ or $R \leq Z_{\alpha'-2}$, and we are done in both cases. □

(9.9) $K = K_\alpha \times K_\beta$; <u>in particular</u> K <u>is abelian</u>.

<u>Proof</u>: Note that $K = K_\alpha K_\beta$. By (9.4) Z_α is a natural module and we have $C_{K_\alpha}(Z_\beta) = 1$. Now the assertion follows.

(9.10) <u>Suppose that</u> $b = 3$, <u>then</u> $q_\alpha = q_\beta = 2$.

<u>Proof</u>: In the previous lemmata we have already shown that $q_\alpha = q_\beta$, $V_{\alpha'} \not\leq Q_\beta$, $L_\delta/Q_\delta \simeq SL_2(q_\alpha)$ for $\delta \in \Gamma$, and that Z_α and V_β/Z_β are natural $SL_2(q_\alpha)$-modules. In addition $Q_\alpha \cap Q_{\alpha+2}$ $(= C_{Q_\alpha}(V_\beta))$ is a normal subgroup of L_β. In the following we assume that $q_\alpha = q_\beta > 2$; i.e. $K \neq 1$.

(1)(a) $[Q_\alpha \cap Q_{\alpha+2}, O^p(L_\beta)] \leq V_\beta$.

(b) $Q_\beta/Q_\alpha \cap Q_{\alpha+2}$ is a natural L_β/Q_β-module.

Note that $[Q_\beta \cap Q_{\alpha+2}, V_{\alpha'}] \leq Z_{\alpha+2} \leq V_\beta \cap V_{\alpha'}$ and by (6.5) $Q_\beta \cap Q_{\alpha+2}$ is not normal in L_β. Since $O^p(L_\beta) \leq \langle V_{\alpha'}^{L_\beta}\rangle$ assertion (a) follows, and since $|Q_\beta/Q_\beta \cap Q_{\alpha+2}| = q_\beta$ the second assertion follows from (9.6) and (5.12).

By (6.7) and (6.8) we may choose $K \leq G_\gamma$. Hence, we may choose $\alpha-1 \in \Delta(\alpha)\setminus\{\beta\}$ and $\alpha'+1 \in \Delta(\alpha')\setminus\{\alpha+2\}$ so that K stabilizes $\alpha-1$ and $\alpha'+1$. In addition, (6.8) and (3.4)(c) imply that $Z_{\alpha'} \not\leq Q_{\alpha-1}$ and $Z_{\alpha-1} \not\leq Q_{\alpha'}$. We define $\hat{\gamma} = (\alpha-1,\alpha,\ldots,\alpha',\alpha'+1)$, $D = C_{Q_{\alpha-1}}(K_{\alpha-1})$, $Q = C_{Q_{\alpha'}}(K)$, $K_\delta = K \cap L_\delta$ and $\langle t_\delta\rangle = K \cap L_\delta \cap G_\delta^{(1)}$ for $\delta \in \hat{\gamma}$.

(2) q_α is even and $K_{\alpha-1} = K \cap G_{\alpha'}^{(1)}$.

Since $K_{\alpha-1}$ centralizes $K_{\alpha'}$ by (9.9) and $Z_{\alpha-1} \not\leq Q_{\alpha'}$, we get from (5.1) that $K_{\alpha-1} \leq G_{\alpha'}^{(1)}$.

Assume that q_α is odd. By (6.9) we have $t_\alpha = t_{\alpha'}$. But this contradicts $Z_\alpha Q_{\alpha'} \in Syl_p(L_{\alpha'})$ and $[Z_\alpha, t_\alpha] \not\leq Q_{\alpha'}$ by (9.4)(a).

Hence, q_α is even and $K_{\alpha'} \cap G_{\alpha'}^{(1)} = 1$. We conclude that $K = K_{\alpha'} \times K_{\alpha-1}$ and $K_{\alpha-1} = K \cap G_{\alpha'}^{(1)}$.

(3) $K_{\alpha-1} \cap G_\delta^{(1)} = 1$ for $\delta \in \{\alpha, \beta, \alpha+2\}$.

From (2) and (3.2)(c) we get $K_{\alpha-1} \cap G_{\alpha+2}^{(1)} = 1$, and (1)(b) (applied to $\alpha-1$) yields $D \leq G_{\alpha-1}^{(2)}$ and thus $K_{\alpha-1} \cap G_\alpha^{(1)} = 1$.

Assume now that $k \in K_{\alpha-1} \cap G_\beta^{(1)}$. Then by (2) $k \in G_\beta^{(1)} \cap G_{\alpha'}^{(1)}$ and, as $[Z_{\alpha-1}, k] = 1$, $[V_\beta, k] \leq Z_\beta$, and there exists $g \in C_{G_{\alpha+2}}(k)$ with $\beta^g = \alpha'$; we conclude that $[V_{\alpha'}, k] \leq Z_{\alpha'}$. It follows that $[Z_{\alpha+2}, k] \leq Z_\beta \cap Z_{\alpha'} = 1$ and $k \in G_{\alpha+2}^{(1)} \cap K_{\alpha-1} = 1$.

(4) $D = Q \times Z_{\alpha-1}$ and $Q_{\alpha+2} \cap Q_{\alpha'+1} = Q \times V_{\alpha'}$; in particular Q is elementary abelian.

Set $H = Q_{\alpha+2} \cap Q_{\alpha'+1}$. An application of (5.1) and (3) shows that $D \leq Q_\delta$ for $\delta \in \hat{\gamma}$, $\delta \neq \alpha'$, and (2) implies that $D = Z_{\alpha-1} \times (D \cap Q_{\alpha'})$ and $D \cap Q_{\alpha'} \leq H$. In addition, we know that $K_{\alpha-1}$ operates fixed-point-freely on $Z_{\alpha+2} = Z_\beta Z_{\alpha'}$ and $V_{\alpha'}/V_{\alpha'} \cap Q_\beta$; i.e. $D \cap V_{\alpha'} = 1$. Since $K_{\alpha'}$ normalizes $D \cap H$ and by (1) $[H, K_{\alpha'}] \leq V_{\alpha'}$, we get $D \cap H \leq C_H(K_{\alpha'})$ and $D \cap H = Q$. Since α' is conjugate to $\alpha-1$, a comparison of the orders of $C_{Q_{\alpha'}}(K_{\alpha'})$ and $D \cap H$ shows that $H = Q \times V_{\alpha'}$; in particular $\phi(H) = \phi(Q)$ and $\phi(Q) \cap Z_{\alpha'} = 1$. As $\phi(Q)$ is normal in $L_{\alpha'}$, this yields $\phi(Q) = 1$.

(5) $Q = 1$.

As we have seen above $Q \leq Q_\delta$ for $\delta \in \hat{\gamma}$ and $Q \leq G_\delta^{(2)}$ for $\delta \in \{\alpha-1, \beta, \alpha'\}$ by (1). It follows that $Q_\alpha \cap Q_{\alpha+2} = Q \times V_\beta$ and $[Q, \langle V_{\alpha-1}, V_{\alpha'}\rangle] = 1$. Now again (1) implies $[Q, Q_\beta] \leq V_\beta$.

From (4) and (1) we get that $D \leq Q_\alpha \cap Q_{\alpha+2}$ and from (3) that $Q_\beta = (Q_\alpha \cap Q_{\alpha+2})[Q_\beta, K_{\alpha-1}]$ and

$$[Q_\beta, Q] = [K_{\alpha-1}, Q_\beta, Q].$$

Now the Three-subgroup Lemma implies that $[Q_\beta, Q] \leq [V_\beta, K_{\alpha-1}]Z_\beta$, and since $V_\beta = Z_{\alpha-1}Z_{\alpha+2}$ we have $[Q_\beta, Q] \leq Z_{\alpha+2}$. On the other hand,

$$[Q_\beta, Q_\alpha \cap Q_{\alpha+2}] = [Q_\beta, QV_\beta] = Z_\beta[Q_\beta, Q]$$

is normal in L_β. We conclude that $[Q, Q_\beta] \leq Z_\beta$, and another application of the Three-subgroup Lemma to Q, Q_β and $\langle V_{\alpha-1}, V_{\alpha'}\rangle$ shows that Q is central in $Q_\beta\langle V_{\alpha-1}, V_{\alpha'}\rangle = L_\beta$. Now (5) follows from $Q \cap Z_\beta = 1$.

(6) $[Q_{\alpha'}, Q_{\alpha'}] \leq Z_{\alpha'}$ and $Q_{\alpha'}/Z_{\alpha'}$ is the direct product of two natural $L_{\alpha'}/Q_{\alpha'}$-modules.

Note that, by (4) and (5), $Q_{\alpha+2} \cap Q_{\alpha'+1} = V_{\alpha'}$. Hence (1) implies that $Q_{\alpha'}/V_{\alpha'}$ and $V_{\alpha'}/Z_{\alpha'}$ are both natural $L_{\alpha'}/Q_{\alpha'}$-modules. Now the assertion follows from (5.17).

We now derive a contradiction. Let $W_{\alpha'} \leq Q_{\alpha'}$ so that $W_{\alpha'} \cap V_{\alpha'} = Z_{\alpha'}$ and $W_{\alpha'}/Z_{\alpha'}$ is a natural $SL_2(q_\alpha)$-module for $L_{\alpha'}/Q_{\alpha'}$. Let $g \in L_{\alpha+2}$ which interchanges the vertices α' and β. Set $W_\beta = W_{\alpha'}^g$, and $M = (W_\beta \cap Q_{\alpha+2})(W_{\alpha'} \cap Q_{\alpha+2})$. Since $Q_{\alpha'} = V_{\alpha'}W_{\alpha'}$ we have $W_{\alpha'} \not\leq Q_{\alpha+2}$. Hence $[M,W_{\alpha'}] \leq W_{\alpha'} \cap Q_{\alpha+2} \leq M$ and M is normal in $L_{\alpha+2}$ and $|M| = q_\alpha^4$. Set $R = C_{Q_{\alpha+2}}(K_{\alpha+2})$.

Since $[Q_{\alpha'},Q_{\alpha'}] \leq Z_{\alpha'}$ we get $[Q_\beta \cap Q_{\alpha'},Q_{\alpha'}] \leq Z_{\alpha'} \leq Z_{\alpha+2}$ and $Q_\beta \cap Q_{\alpha'} = RZ_{\alpha+2}$ is normal in $L_{\alpha+2}$, and $M/Z_{\alpha+2}$ is a natural $SL_2(q_\alpha)$-module for $L_{\alpha+2}/Q_{\alpha+2}$.

If $M \cap Q_{\alpha'} \leq Q_{\alpha'+1}$, then $M \cap Q_{\alpha'} = Q_{\alpha+2} \cap Q_{\alpha'+1} = V_{\alpha'}$ which contradicts $M \cap W_{\alpha'} \not\leq V_{\alpha'}$. Hence $M \cap Q_{\alpha'} \not\leq Q_{\alpha'+1}$ and $K_{\alpha+2} \not\leq G_{\alpha'+1}^{(1)}$. As $C_M(K_{\alpha+2}) = 1$ we have that $R \leq Q_{\alpha+2} \cap Q_{\alpha'+1}$ and $V_{\alpha'} = RZ_{\alpha+2}$, a contradiction to (3.2)(c). □

(9.11) <u>Suppose that</u> $b > 3$, <u>then</u> $b = 5$ <u>and</u> $q_\alpha = 3$.

<u>Proof:</u> Assume first that $b > 5$. By (9.7) there exists $\alpha-1 \in \Delta(\alpha)$ and $\alpha-2 \in \Delta(\alpha-1)$ such that $Z_{\alpha-2} \not\leq Q_{\alpha'-1}$. It follows with (8.4) that the hypotheses of this chapter are also fulfilled with $\alpha-2$ and $\alpha'-2$ in place of α and α'. Hence we have by (9.7) that $V_{\alpha-2}^{(2)} \not\leq Q_{\alpha'-4}$. Since $b \geq 7$, we get that $[R,V_{\alpha-2}^{(2)}] = 1$ for $R = [V_\beta,V_{\alpha'}]$. Again by (9.7) and (9.8) it follows that $R = C_{Z_{\alpha'-1}}(V_\alpha^{(2)}) = Z_{\alpha'-2}$. Thus $V_{\alpha-2}^{(2)}$ centralizes $Z_{\alpha'-2}Z_{\alpha'-4} = Z_{\alpha'-3}$ and $Z_{\alpha'-3}$ is normal in $\langle V_{\alpha-2}^{(2)},G_{\alpha'-3} \cap G_{\alpha'-4}\rangle = G_{\alpha'-4}$, a contradiction to (3.2)(c). This contradiction shows that $b = 5$.

Let $(\delta_o,\ldots,\delta_4)$ be any path of length 4 in Γ with $\delta_o \in \beta^G$. If $\delta_o = \beta$ and $\delta_4 = \alpha'$, then (9.8) and (9.4)(a) imply that

(*) $$[V_{\delta_o}, V_{\delta_4}] = Z_{\delta_2},$$

and since $s \geq 4$ by (6.8), this statement holds for every such path.

Assume next that q_α is even. Let $\delta \in \Gamma$ with $(\alpha+3,\alpha+4,\alpha',\ldots,\delta)$ of length 5. Then as $b = 5$ and $s \geq 4$ we may assume without loss that $Z_\delta \nleq Q_{\alpha+3}$ and we can choose $t \in Z_\alpha \setminus Q_{\alpha'}$ and $\tilde{t} \in Z_\delta \setminus Q_{\alpha+3}$. Set $z = [t^{\tilde{t}}, t]$ and $\tilde{z} = [\tilde{t}^t, \tilde{t}]$. It follows from (*) that

$$z \in [Z_\alpha, Z_\alpha^{\tilde{t}}] \leq Z_{\alpha+3} \quad\text{and}\quad \tilde{z} \in [Z_\delta, Z_\delta^t] \leq Z_{\alpha'}.$$

On the other hand, $\langle t,\tilde{t}\rangle$ is a dihedral group of order $2|\langle t^{\tilde{t}},t\rangle| \leq 16$ and thus $z = (\tilde{t}t)^4 = \tilde{z} = (t\tilde{t})^4$. It follows that $z \in Z_{\alpha+3} \cap Z_{\alpha'}$, and $b = 5$ and (9.4)(a) imply that $z = \tilde{z} = 1$. Since $V_\beta = Z_\alpha Z_{\alpha+2}$ and $[Z_{\alpha+2}, Z_{\alpha+2}^{\tilde{t}}] = 1$ we have shown that $[V_\beta, V_\beta^{\tilde{t}}] = 1$, and since $(\beta^{\tilde{t}},\ldots,\beta)$ is conjugate to $(\beta,\ldots,\alpha')$, we also have $[V_\beta, V_{\alpha'}] = 1$, a contradiction.

We have shown that $q_\alpha (= q_\beta)$ is odd; in particular, $Z(L_\delta/Q_\delta)$ has order 2 for $\delta \in \Gamma$. By (6.7) and (6.8) we may assume that K stabilizes $(\alpha,\ldots,\alpha+4)$. Hence, there exists $\lambda \in \Delta(\alpha+4)\setminus\{\alpha+3\}$ so that $K \leq G_\lambda$. Since the paths $(\beta,\ldots,\alpha')$ and $(\beta,\ldots,\lambda)$ are conjugate, we have $V_\lambda \nleq Q_\beta$, and to avoid additional notation we set $\lambda = \alpha'$. In addition, we define $\langle t_\delta \rangle = L_\delta \cap K \cap G_\delta^{(1)}$ for $\delta \in \gamma$ and $D = C_{Q_{\alpha+3} \cap Q_{\alpha+2}}(t_\beta)$.

By (6.9) we have $t_\beta = t_{\alpha+4}$ and $Q_\beta \cap D = C_{Q_\beta}(t_\beta)$, and by (9.8)

$V_{\alpha'} = (V_{\alpha'} \cap D)Z_{\alpha+4}$. Set $L = \langle (D \cap V_{\alpha'})^x \mid x \in C_{L_\beta}(t_\beta)\rangle$. Then $L/L \cap Q_\beta \simeq SL_2(q_\beta)$ and L operates (possibly non-faithfully) as $L_2(q_\beta)$ on the chief factors of $D \cap Q_\beta$. Hence, by [8,3.8.3] it suffices to show that $D \cap V_{\alpha'}$ operates non-trivially and quadratically on $D \cap Q_\beta$.

If $[D \cap V_{\alpha'}, D \cap Q_\beta] = 1$, then $\langle D, G_{\alpha'} \cap G_{\alpha+4}\rangle = G_{\alpha+4}$ normalizes $Z_{\alpha+4}(D \cap V_{\alpha'}) = V_{\alpha'}$ contradicting (3.2)(c). Note that $[V^{(2)}_{\alpha+4}, V^{(2)}_{\alpha+4}] = 1$ since $b = 5$. Hence, we have $[D \cap Q_\beta, D \cap V_{\alpha'}, D \cap V_{\alpha'}] \leq [V^{(2}_{\alpha+4}, V^{(2)}_{\alpha+4}] = 1$.

□

10. The case $b \leq 2$ and $[Z_\alpha, Z_{\alpha'}] \neq 1$.

In the chapters 7 - 9 we have shown that $b = 1, 2, 3$ or 5. In this chapter we start the analysis of the chief factors of G_δ in Q_δ $(\delta = \alpha, \beta)$. Our goal is to determine these chief factors and the operation of G_δ on them.

In this chapter we assume:

(10.0) Hypothesis. Hypothesis B' holds and furthermore:

(a) $[Z_\alpha, Z_{\alpha'}] \neq 1$,

(b) $b \leq 2$.

We will be able to show that G is parabolic isomorphic to J_2 or $G_2(2)'$ or that $r = s-1$ and stabilizers of paths of length s are p'-groups. In chapter 15 this latter case will lead to the conclusion that G is locally isomorphic to X for $X \in \{L_3(q), Sp_4(q)$ (q even), ${}^3D_4(q), G_2(q)\}$ and appropriately chosen $q = p^n$.

Notation. For this chapter let:

$$\alpha-1 \in \Delta(\alpha) \setminus \{\beta\},$$

$$\alpha'+1 \in \Delta(\alpha') \setminus \{\alpha'-1\},$$

$$D_\alpha = Q_\beta \cap Q_{\alpha-1},$$

$$D_\alpha' = Q_{\alpha'-1} \cap Q_{\alpha'+1},$$

$$K_\delta = K \cap L_\delta, \quad \delta \in \Gamma,$$

$$W_\alpha = \langle Z_{\alpha'}, Z_{\alpha'}^g \rangle, \quad g \in L_\alpha \text{ with } \beta^g = \alpha-1,$$

$$R_\alpha = C_{D_\alpha}(W_\alpha).$$

Since $b \leq 2$ we may assume that K fixes the path $(\alpha-1,\alpha,\ldots,\alpha',\alpha'+1)$. In particular, for $(\delta,\delta+1)$ on this path, $K_\delta T_\delta = N_{L_\delta}(T_\delta)$ where T_δ is a Sylow p-subgroup of $L_\delta \cap L_{\delta+1}$.

The next two lemmata sum up what we have already proven in chapter 7.

(10.1) <u>For each</u> $\delta \in \{\alpha,\alpha'\}$ <u>we have that</u>

(a) $L_\delta/Q_\delta \simeq SL_2(q_\delta)$ <u>and</u> $q_\alpha = q_{\alpha'}$,

(b) $Z_\delta/\Omega_1(Z(L_\delta))$ <u>is a natural</u> L_δ/Q_δ<u>-module</u>,

(c) $Q_\alpha Q_\beta = Q_\alpha Z_{\alpha'} \in Syl_p(L_\alpha)$ <u>and</u> $Q_{\alpha'-1}Q_{\alpha'} = Z_\alpha Q_{\alpha'} \in Syl_p(L_{\alpha'})$.

(10.2) <u>Assume</u> $Z_\beta \leq Z(L_\beta)$. <u>Then</u>

(a) $b = 2$,

(b) $Z_\beta = \Omega_1(Z(L_\beta))$ <u>and</u> $Z(L_\alpha) = 1$,

(c) $Z_\alpha = Z_\beta \times Z_{\alpha-1}$ <u>and</u> Z_α <u>is a natural</u> L_α/Q_α<u>-module</u>,

(d) $[Q_\beta,V_\beta] = Z_\beta$,

(e) $K = K_\alpha \times K_\beta$.

<u>Proof:</u> (10.1) is (7.1), and (10.2) is, apart from (e), (7.4) and (7.5) of chapter 7. It remains to prove (10.2)(e). By definition $K = K_\alpha K_\beta$, and since K_α operates fixed-point-freely on Z_α while K_β centralizes Z_β, we get $K_\alpha \cap K_\beta = 1$. □

(10.3) <u>Assume</u> $Z_\beta \nleq Z(L_\beta)$, <u>then</u> $Z_{\alpha-1} \nleq G_{\alpha'}$. <u>In particular</u>, $(\alpha-1,\alpha'-1)$ <u>has the same properties as</u> (α,α').

Proof: Assume to the contrary that $Z_{\alpha-1} \leq G_{\alpha'}$. By (10.1) we have $Z_{\alpha-1} \leq Z_\alpha Q_{\alpha'}$ and $[Z_{\alpha-1}, Z_{\alpha'}] \leq [Z_\alpha, Z_{\alpha'}] \leq Z_\alpha$. Thus, $Z_\alpha Z_{\alpha-1}$ is normal in $\langle G_\alpha \cap G_{\alpha-1}, Z_{\alpha'} \rangle = G_\alpha$, contradicting (6.6). □

(10.4) For $\delta \in \{\alpha, \alpha'\}$ the following hold:

(a) $\phi(D_\delta) = 1$,

(b) D_δ is normal in L_δ,

(c) $D_\alpha = R_\alpha$ and $b = 1$; or $D_\alpha = R_\alpha Z_\alpha$, $R_\alpha = D_\alpha \cap Q_{\alpha'} \cap Q_\alpha^g$, and $b = 2$.

Proof: It follows from (10.1) that $W_\alpha Q_\alpha = L_\alpha$.

If $b = 1$, then $R_\alpha = D_\alpha$, and D_α is normal in L_α. If $b = 2$, then (10.1) implies that

$$[Z_\alpha, Z_{\alpha'}] = Z_\alpha \cap Z_{\alpha'} \quad \text{and} \quad [D_\alpha, Z_{\alpha'}] \leq Z_\alpha \leq D_\alpha.$$

With the same argument $[D_\alpha, Z_\alpha^g] \leq D_\alpha$, and again D_α is normal in $W_\alpha Q_\alpha = L_\alpha$. With the roles of α and α' reversed we get that $D_{\alpha'}$ is normal in $L_{\alpha'}$, and so (b) holds. Note that this implies that $\phi(D_\delta)$ is normal in L_δ.

If $b = 1$, then (10.1) implies that

$$Q_\alpha \cap Q_\beta = D_\alpha(Z_\alpha \cap Q_\beta) = D_\beta(Z_\beta \cap Q_\alpha)$$

and thus

$$\phi(Q_\alpha \cap Q_\beta) = \phi(D_\alpha) = \phi(D_\beta).$$

Now (3.2)(c) yields that $\phi(D_\alpha) = \phi(D_\beta) = 1$.

If $b = 2$, then $R_\alpha = D_\alpha \cap Q_{\alpha'} \cap Q^g_{\alpha'}$, and we get that $D_\alpha = R_\alpha Z_\alpha$ and $\phi(D_\alpha) = \phi(R_\alpha)$. Assume that $Z_\beta \leq Z(L_\beta)$, then by (10.2) we get $R_\alpha \cap Z_\alpha = 1$ and so $\Omega_1(Z(S)) \cap R_\alpha = 1$. It follows that $\phi(D_\alpha) = 1$. Assume that $Z_\beta \nleq Z(L_\beta)$, then by (10.3) (with the corresponding notation) $R_\beta = D_\beta \cap Q_{\alpha-1} \cap Q_{\alpha'+1}$. Since $R_\alpha \leq Q_{\alpha'}$, we get $\phi(R_\alpha) \leq R_\beta$, and (3.2)(c) yields $\phi(R_\alpha) = \phi(D_\alpha) = 1$. □

(10.5) Assume $b = 1$. Then for each $\delta \in \{\alpha,\beta\}$:

(a) $L_\delta/Q_\delta \simeq SL_2(q_\delta)$, $q_\alpha = q_\beta$.

(b) $Q_\delta = Z_\delta$ and $Q_\delta/Z(L_\delta)$ is a natural L_δ/Q_δ-module.

(c) $r = s-1$, and either

(i) $r = 3$ and $Z(L_\delta) = 1$, or

(ii) $r = 4$, $|Z(L_\delta)| = q_\delta$ and q_δ is even.

(d) The stabiliers of paths of length s are p'-groups.

Proof: Note that $\alpha' = \beta$. Thus (a) is (10.1).

Set $q = q_\delta$; again by (10.1) we have $Q_\delta = Z_\delta D_\delta$ and $|Q_\delta/D_\delta| = q^2$ and by (10.4)(c) $D_\delta \leq Z(L_\delta)$. Together with (10.4)(a) it follows that $Q_\delta = Z_\delta$, and (10.1) implies (b).

We have $D_\delta \leq Z(L_\delta)$ and clearly also $Z(L_\delta) \leq D_\delta$ since Q_α and $Q_{\alpha'}$ contain their centralizers in L_α, resp. $L_{\alpha'}$. Now an easy application of (3.2)(c) shows that $D_\alpha \cap D_\beta = 1$ and the action of K on D_δ implies that either $|D_\delta| = 1$ or $|D_\delta| = q$.

We have shown that $|S| = q^3$ and $Z(L_\delta) = 1$ or $|S| = q^4$ and $|Z(L_\delta)| = q$. It is now easy to check, with the help of (3.4)(c), that $r = s-1 = 3$ in the first case and $r = s-1 = 4$ in the second case. In both cases (d) holds.

We need then only show that q is even if $|Z(L_\delta)| = q$. So suppose that q is odd. Let $t_\alpha, t_\beta, t_{\alpha'+1}$ be defined as in (6.9). Note that $Z(L_\alpha) \not\leq Q_{\alpha'+1}$. Hence, the action of K implies that $Z(L_\alpha)Q_{\alpha'+1} \in Syl_p(L_{\alpha'+1})$. It follows that $t_\alpha \in G_{\alpha'+1}^{(1)}$. On the other hand, (6.9) yields $t_\alpha \neq t_{\alpha'+1}$ and $t_\beta \in \langle t_\alpha, t_{\alpha'+1}\rangle \leq G_{\alpha'+1}^{(1)}$. But this contradicts (3.2)(c). □

(10.6) <u>Assume</u> $Z_\beta \not\leq Z(L_\beta)$ <u>and</u> $b = 2$. <u>Then for each</u> $\delta \in \{\alpha,\beta\}$:

(a) $L_\delta/Q_\delta \simeq SL_2(q_\delta)$.

(b) $|Q_\delta| = q_\delta^5$, $\phi(Q_\delta) = Z(L_\delta)$, <u>and</u> Q_δ/Z_δ <u>and</u> $Z_\delta/Z(L_\delta)$ <u>are natural</u> L_δ/Q_δ<u>-modules</u>.

(c) $q_\alpha = q_\beta = 3^n$, <u>for some</u> $n \in \mathbb{N}$.

(d) $r = s-1 = 6$.

(e) <u>The stabilizers of paths of length</u> s <u>are</u> p'<u>-groups</u>.

<u>Proof</u>: It follows from (10.3) and the minimality of b that $b_\alpha = b_\beta = b$. Thus we may reverse the roles of α and β so far; in particular, (10.1) implies that (a) holds.

We have that $Q_\alpha = Z_{\alpha-1}Z_\beta D_\alpha$, and (10.4) implies that $D_\alpha = Z_\alpha$ and $R_\alpha = Z(L_\alpha)$. With the same argument $D_\beta = Q_\alpha \cap Q_{\alpha'} = Z_\beta$. In addition,

$$[Q_\alpha, Z_{\alpha'}, Z_{\alpha'}] \leq [Q_\alpha \cap Q_\beta, Z_{\alpha'}] = [Z_\alpha, Z_{\alpha'}] \leq Z_\alpha,$$

while similarly

$$[Q_\beta, Z_{\alpha-1}, Z_{\alpha-1}] \leq Z_\beta.$$

Since $|Q_\alpha/Z_\alpha| = q_\beta^2$ and $|Q_\beta/Z_\beta| = q_\alpha^2$ we get with (5.12) that

$q_\alpha = q_\beta$ and Q_δ/Z_δ and $Z_\delta/Z(L_\delta)$ are natural L_δ/Q_δ-modules. In particular, the non-trivial elements in Q_δ/Z_δ are conjugate under L_δ.

If q_δ is even, then every element in $Q_\alpha \setminus Z_\alpha$ is an involution since $Z_{\alpha-1} \leq Q_\alpha$ but $Z_{\alpha-1} \nleq Z_\alpha$. Hence, Q_α is abelian and $[Z_{\alpha-1}, Z_\beta] = 1$, contradicting (10.3).

Thus, q_δ is odd. Set $U = [Z_\alpha, Z_{\alpha'}]$. Then U is centralized by $\langle Q_\alpha, Q_{\alpha'}\rangle = L_\beta$. The module structure of Z_α and an easy application of (3.2)(c) yield

$$U = Z(L_\beta),\ |U| = q_\delta \text{ and } |Q_\delta| = q_\delta^5.$$

The assertions (d) and (e) are now easy consequences, and (c) follows from [13, 3.4]. □

(10.7) <u>Assume that</u> $Z_\beta \leq Z(L_\beta)$, <u>then for</u> $\bar{Q}_\alpha = Q_\alpha/D_\alpha$:

(a) $\bar{Q}_\alpha$ <u>is elementary abelian,</u>

(b) $\bar{Q}_\alpha = (\overline{V_\beta \cap Q_\alpha}) \times (\overline{V_{\alpha-1} \cap Q_\alpha})$, <u>and</u> $|\bar{Q}_\alpha| = q_\beta^2$ <u>and</u> $q_\beta \geq q_\alpha$,

(c) $Q_\beta = V_\beta D_\alpha$ <u>and</u> $\phi(Q_\beta) = Z_\beta$,

(d) $L_\beta/Q_\beta \simeq SL_2(q_\beta)$ <u>or</u> $L_2(q_\beta)$.

<u>Proof:</u> Note that by (10.4) D_α is normal in L_α and, since $V_\beta \cap D_\alpha$ is normalized by W_α, $V_\beta \cap D_\alpha = V_\beta \cap V_{\alpha-1}$.

Set $V = (V_\beta \cap Q_\alpha)(V_{\alpha-1} \cap Q_\alpha)D_\alpha$. If $[\bar{V}, O^p(L_\alpha)] = 1$, then $[Q_\alpha, O^p(L_\alpha)] \leq V_\beta \cap V_{\alpha-1}$ and $Q_\alpha \cap Q_\beta$ is normal in L_α, contradicting (6.5). Thus $[\bar{V}, O^p(L_\alpha)] \neq 1$.

Since $V'_\lambda = Z_\lambda \leq D_\alpha$ for $\lambda \in \Delta(\alpha)$, we have that $\bar{V}$ is elementary abelian. It follows that $\bar{V}$ is a KW_α-module and

$$\bar{V} = (\overline{V_{\alpha-1} \cap Q_\alpha}) \times (\overline{V_\beta \cap Q_\alpha}).$$

In addition, V_β operates quadratically on $\bar{V}$, and (5.12) implies $q_\alpha \leq |\overline{V_{\alpha-1} \cap Q_\alpha}|$.

Assume first that q_α is odd. Then there exists an involution $t_\alpha \in K_\alpha \cap G_\alpha^{(1)}$ and by (5.9)(a)

$$\bar{Q}_\alpha = C_{\bar{Q}_\alpha}(t_\alpha) \times \bar{V}.$$

In particular, $C_{\bar{Q}_\alpha}(t_\alpha) \cap \bar{Q}_\beta = 1$ and $S/Q_\beta = C_{S/Q_\beta}(t_\alpha) \times VQ_\beta/Q_\beta$. Now (5.1) implies $S = VQ_\beta$ and $Q_\alpha = V$.

Assume now that q_α is even. By (5.11) together with (5.4)(d) we have $[\bar{V},K_\alpha] = \bar{V}$ or $K_\alpha = 1$. In the first case $\bar{Q}_\alpha = C_{\bar{Q}_\alpha}(K_\alpha) \times \bar{V}$, and we get as above $C_{\bar{Q}_\alpha}(K_\alpha) = 1$. In the second case $L_\alpha/Q_\alpha \simeq SL_2(2)$ and $\bar{Q}_\alpha = C_{\bar{Q}_\alpha}(O^2(L_\alpha)) \times \bar{V}$, and again as above $C_{\bar{Q}_\alpha}(O^2(L_\alpha)) = 1$.

Assume now that $V = Q_\alpha$. Then (a) holds, and S/Q_β is elementary abelian. Hence (d) holds, too, and $S = (V_{\alpha-1} \cap Q_\alpha)Q_\beta$. This implies (b) and together with (10.2) and (10.4) also (c). □

(10.8) <u>Assume</u> $Z_\beta \leq Z(L_\beta)$, <u>then</u> $R_\alpha \cap Q_{\alpha'+1} = 1$.

<u>Proof</u>: Set $Q = R_\alpha \cap Q_{\alpha'+1}$. Note that $Q \leq D_{\alpha'}$ and note further that W_α is transitive on $\Delta(\alpha)$ and by (10.7) $q_\alpha \leq q_\beta$. Since $[Z_\alpha,Q] = 1$ we get from (10.2)(c) (applied to α') $Q \cap Z_{\alpha'} = Q \cap Z_\beta$ and from (3.2)(c)

$$Q \cap Z_{\alpha'} = Q \cap Z_\beta = 1.$$

The structure of L_α/Q_α shows that we may choose $g \in N_{L_\alpha}(K)$ so that $\beta^g = \alpha-1$. Hence by the definition of W_α we may assume that K normalizes W_α and $K_\alpha \leq W_\alpha$; in particular $K \leq N_G(Q)$. Since $Q \leq D_{\alpha'}$ and by (10.4)(c) (applied to α') $[Q, O^p(L_{\alpha'})] \leq Z_{\alpha'}$, we get $[Q, K_{\alpha'}] \leq Q \cap Z_{\alpha'} = 1$.

Assume first that $K_\alpha \neq 1$ (and thus $K_{\alpha'} \neq 1$). Choose $t \in N_{L_\beta}(K)$ with $\alpha^t = \alpha'$; i.e. $K_\alpha^t = K_{\alpha'}$. If $K_\alpha^t = K_\alpha$, then by (10.2)(e) $[K_\alpha, t] \leq K_\alpha \cap K_\beta = 1$, and the structure of $\mathrm{Aut}(L_\beta/Q_\beta)$ implies $[K_\alpha, L_\beta] \leq Q_\beta$; in particular $[K_\alpha, V_{\alpha-1} \cap Q_\alpha] \leq D_\alpha$. Applying (10.7) and the action of g we get $[K_\alpha, Q_\alpha] \leq D_\alpha$. But then either L_α centralizes Q_α/D_α or $K_\alpha \leq G_\alpha^{(1)} \cap L_\alpha$. In the first case $Q_\alpha \cap Q_\beta$ is normal in G_α, contradicting (6.5), and the second case contradicts (3.2)(c) applied to $C_G(K_\alpha)$ since $[L_\beta, K_\alpha] \leq Q_\beta$.

Hence, $K_\alpha \neq K_\alpha^t$ and $H = K_\alpha K_\alpha^t \cap K_\beta \neq 1$. As $[Q, K_{\alpha'}] = 1$ we have $[Q, H] = 1$, and as $[Q, V_\beta \cap Q_\alpha] \leq Z_\beta$ we have

$$[Q, V_\beta \cap Q_\alpha, H] = [H, Q, V_\beta \cap Q_\alpha] = 1.$$

Now the Three-subgroup Lemma yields

$$[V_\beta \cap Q_\alpha, H, Q] = 1.$$

Suppose that $[V_\beta \cap Q_\alpha, H] \leq D_\alpha$. Since $(V_\beta \cap Q_\alpha)Q_{\alpha-1} \in \mathrm{Syl}_p(L_{\alpha-1})$ and $Z_\beta Q_{\alpha-2} \in \mathrm{Syl}_p(L_{\alpha-2})$ for $\alpha-2 \in \Delta(\alpha-1) \setminus \{\alpha\}$, it follows by (5.1) that

$[L_{\alpha-1},H] \leq Q_{\alpha-1}$ and $[L_{\alpha-2},H] \leq Q_{\alpha-2}$ contradicting (3.2)(c).

Thus, we have $[V_\beta \cap Q_\alpha,H] \nleq D_\alpha$ and the action of K yields $[V_\beta \cap Q_\alpha,H](V_\beta \cap D_\alpha) = V_\beta \cap Q_\alpha$. Now by (10.4)(a) $V_\beta \cap Q_\alpha$ centralizes Q, and the action of W_α and (10.7)(b) gives $Q \leq Z(L_\alpha)$ contradicting (10.2)(b).

Assume next that $K_\alpha = 1$ but $K_\beta \neq 1$, i.e. $L_\alpha/Q_\alpha \simeq \Sigma_3$ and $q_\alpha = 2$ but $q_\beta \geq 4$. Let $1 \neq x \in Q$. Since $[V_{\alpha-1},x] \leq Z_{\alpha-1}$, we have from (10.2)(c) that $|V_{\alpha-1}/C_{V_{\alpha-1}}(x)| \leq 2$, and since $x \in D_{\alpha'}$, we get with the same argument $|V_\lambda/C_{V_\lambda}(x)| \leq 2$ for $\lambda \in \Delta(\alpha')$.

Set $N_\beta = \langle C_{V_{\alpha-1} \cap Q_\alpha}(x), C_{V_{\alpha'+1} \cap Q_{\alpha'}}(x)\rangle$. As a consequence of (3.2)(c) we know that N_β is not transitive on $\Delta(\beta)$. Thus, by (10.7) and (5.2) we have

$$N_\beta Q_\beta/Q_\beta \simeq \Sigma_3 \quad \text{and} \quad L_\beta/Q_\beta \simeq SL_2(4).$$

Let d be an element of order 3 in N_β. Then there exists $\alpha'' \in \Delta(\beta)\setminus\{\alpha\}$ so that $d \in L_\beta \cap G_{\alpha''}$. On the other hand, since $L_{\alpha''}/Q_{\alpha''} \simeq \Sigma_3$, we have $d \in G_{\alpha''}^{(1)}$, and by (10.7) d operates fixed-point-freely on $Q_{\alpha''}/D_{\alpha''}$ where $D_{\alpha''} = Q_\beta \cap Q_\rho$, $\rho \in \Delta(\alpha'')\setminus\{\beta\}$; in particular $x \in D_{\alpha''}$. It follows as above that $|V_\rho/C_{V_\rho \cap Q_{\alpha''}}(x)| \leq 2$ and $V_\rho \cap Q_{\alpha''} \nleq Q_\beta$. Since d normalizes $V_\rho \cap Q_{\alpha''}$ we get that $\langle N_\beta,V_\rho \cap Q_{\alpha''}\rangle$ is transitive on $\Delta(\beta)$ contradicting (3.2)(c).

Assume now that $L_\alpha/Q_\alpha \simeq \Sigma_3 \simeq L_\beta/Q_\beta$. Then $[V_\beta \cap Q_\alpha,D_\alpha] \leq Z_\beta$ and so $|D_\alpha/C_{D_\alpha}(V_\beta \cap Q_\alpha)| \leq 2$, and $|R_\alpha/C_{R_\alpha}(V_\beta \cap Q_\alpha)| \leq 2$. Since W_α operates transitively on $\Delta(\alpha)$ and by (10.2)(b), $|R_\alpha| \leq 2$ so $2^4 \leq |Q_\beta| \leq 2^5$. Thus we may assume that $|Q_\beta| = 2^5$, $Q = R_\alpha$ and $C_{Q_\alpha}(Q) = Z_\alpha Q$.

Clearly $|V_\beta| \leq 2^4$ and the operation of L_β implies that either

$V'_\beta = 1$ or $V_\beta \simeq C_4 * Q_8$. The first possibility contradicts $[Z_\alpha, Z_{\alpha'}] \neq 1$, so the second holds. Thus Q inverts $Z(V_\beta)$, since $[Q,V_\beta] = 1$ would imply $Q \leq Z_\beta$.

Set $D = QZ(V_\beta)$ and pick $t \in L_\beta$ so that $\alpha^t = \alpha'$. Since all involutions in $D_{\alpha'} \setminus Z_{\alpha'}$ are conjugate under the operation of $Q_{\alpha'}$ we may further choose t so that $Q^t = Q$ and so $D^t = D$. As t centralizes QZ_β and $D \simeq D_8$, t induces an inner automorphism on B. Thus, there exists $t' \in L_\beta$ with $t' \notin Q_\beta$ and $[D,t'] = 1$. As $Q \leq D_{\alpha'}$ and $Z_{\alpha'+1} \leq V_\beta$ we conclude that $Z_{\alpha'+1}$ centralizes D. Hence replacing t' by a suitable element of $t'Z_{\alpha'+1}$ and noting that $Z_{\alpha'+1} \not\leq Q_\delta$, for $\delta \in \Delta(\beta) \setminus \{\alpha,\alpha'\}$, we may assume that $t' \in Q_\delta$. If $Q \leq Q_\delta$, then Q is centralized by V_β, which is not the case. Hence $Q \not\leq Q_\delta$ and also $D \not\leq Q_\delta$. But the structure of Q_δ now implies that $t' \in C_{Q_\delta}(D) \leq Q_\delta \cap Q_\beta$, a contradiction. □

(10.9) <u>Assume that</u> $Z_\beta \leq Z(L_\beta)$ <u>and</u> $D_\alpha = Z_\alpha$. <u>Then</u> $r = 6$, $s = 5$ <u>and</u> G <u>is parabolic isomorphic to</u> $G_2(2)'$ <u>or</u> J_2.

<u>Proof</u>: Note that (10.7) yields $|S| = q_\alpha^3 q_\beta^2$. In addition, we have $Q_\beta = (Q_\alpha \cap Q_{\alpha'})Z_\alpha Z_{\alpha'}$ and $L_\beta = \langle Q_\alpha, Q_{\alpha'}\rangle Q_\beta$.

Assume first that q_α is odd. Then $K_\alpha \cap G_\alpha^{(1)} = \langle t_\alpha\rangle \simeq C_2$ since $L_\alpha/Q_\alpha \simeq SL_2(q_\alpha)$. Note that K is abelian. By (3.2)(c) $[t_\alpha, L_\beta] \not\leq Q_\beta$ and thus by (5.1) t_α operates fixed-point-freely on Q_α. Hence, Q_α is abelian and Q_α operates quadratically on Q_β. It follows from (10.7)(d) and (5.9)(a) that $L_\beta/Q_\beta \simeq SL_2(q_\beta)$, and there exists an involution $t_\beta \in K_\beta \cap G_\beta^{(1)}$. Now (6.9) implies that

$$q_\alpha = |C_S(t_\alpha)| = |C_S(t_\beta)| \geq q_\beta |Z_\beta|$$

contradicting (10.7)(b).

Thus, we have proven that q_α is even. Assume next that Q_α is elementary abelian. Then $Q_\alpha \cap Q_{\alpha'} = Z_\beta$ and $|Q_\beta/Z_\beta| = q_\alpha^2$ contradicting $|S| = q_\beta^2 q_\alpha^3$.

By (10.7) and (5.11) an element of order 3 in L_α acts fixed-point-freely on Q_α. If $q_\alpha > 2$, then by (5.17) Q_α is elementary abelian, a contradiction. Thus, $q_\alpha = 2$, $|Q_\beta/Z_\beta| = 4q_\beta$, and (5.7) leads to $q_\beta \leq 4$.

The case $q_\beta = 4$: Then Q_α is special of order 2^6 and Q_β is extra-special of order 2^5, and (5.14) implies that Q_β/Z_β is an orthogonal L_β/Q_β-module. In addition, $K_\beta \leq G_\alpha^{(1)}$ since $q_\alpha = 2$, and so K_β normalizes $V_\mu \cap Q_\alpha$, $\mu \in \Delta(\alpha)$. Hence, as $V_\mu = Q_\mu$ (10.7) yields that there exist two K_β-invariant complements for Q_β in S. We conclude with (5.16) that L_β is isomorphic to the semi-direct product of $SL_2(4)$ ($\simeq A_5$) and an extra-special group of order 2^5 which involves an orthogonal $SL_2(4)$-module. We leave it as an easy exercise to show that G is parabolic isomorphic to J_2 (see also chapter 15).

Let $\tilde{\gamma} = (\alpha,\ldots,\alpha+6)$ be a path of length 6 in Γ. Then $O_2(G_{\tilde{\gamma}}) = Z_{\alpha+3}$ and $r = 6$. The order of S shows that $s = 5$.

The case $q_\beta = 2$: We have $|Q_\alpha| = |Q_\beta| = 2^4$, and (10.7) yields $Q_\alpha' = 1$ and, as seen above, $Q_\alpha \simeq C_4 \times C_4$ and $Q_\alpha \cap Q_{\alpha'} = Z(Q_\beta) \simeq C_4$. It follows that $Q_\beta \simeq C_4 * Q_8$ and it is again an easy exercise to show that G is parabolic isomorphic to $G_2(2)'$. Furthermore, just as above, it follows that $r = 6$, $s = 5$. □

(10.10) Assume that $Z_\beta \leq Z(L_\beta)$ and $D_\alpha \nleq Z_\alpha$, then the following hold:

(a) $r = s-1 = 6$.

(b) $|D_\alpha| = q_\alpha^2 q_\beta$ and $|S| = q_\alpha^3 q_\beta^3$.

(c) The stabilizers of paths of length s are p'-groups.

(d) $\phi(Q_\beta) = Z_\beta$, and either Q_β/Z_β is an irreducible G_β/Q_β-module, or $q_\alpha = q_\beta = 2$ and $[V_\beta,L_\beta] \simeq Q_8$.

Proof: As in (10.8) we may choose K to normalize W_α and stabilize $\tilde{\gamma} = ((\alpha'+1)^g,\ldots,\alpha,\ldots,\alpha'+1)$, g as in the definition of W_α. Hence, we have by (10.4) and (10.7) $G_{\tilde{\gamma}} = KR_\alpha$.

Since (10.8) shows that $R_\alpha \cap Q_{\alpha'+1} = 1$ and symmetrically $R_\alpha \cap Q^g_{\alpha'+1} = 1$, we get from (10.7)(d) and the action of K on R_α that $|R_\alpha| = q_{\alpha'+1}$; in particular $\tilde{\gamma}$ is regular and $|S| = q_\alpha^3 q_\beta^3$, and (a), (b) and (c) follow.

It remains to prove (d). If $q_\beta = 2$, then by (10.7) $q_\alpha = 2$ and so $|V_\beta| \leq 2^4$. Since $[Z_\alpha,Z_{\alpha'}] \neq 1$ it follows by (10.7) that $|V_\beta| = 2^4$, $V_\beta \simeq C_4 * Q_8$ and $[V_\beta,L_\beta] \simeq Q_8$.

Assume now that $q_\beta > 2$. Pick $x \in N_{L_\beta}(K_\beta) \setminus K_\beta$. Then $Q_\beta = R_\alpha R_\alpha^x Z_\alpha Z_{\alpha'}$ and as $[K_\alpha,R_\alpha] = 1$ the operation of K on R_α implies $C_{Q_\beta}(K_\beta) \leq Z_\alpha Z_{\alpha'}$, and we have $[Q_\alpha,K_\beta]D_\alpha = Q_\alpha$ and $[Q_\alpha,K_\beta,D_\alpha] = Z_\alpha$. Similarly $[Q_{\alpha'},K_\beta,D_{\alpha'}] = Z_{\alpha'}$, and we get

$$[Q_\beta,L_\beta] \geq [Q_\beta,K_\beta]Z_\alpha Z_{\alpha'} = Q_\beta.$$

Therefore, $Q_\beta = [Q_\beta,L_\beta]$ and $L_\beta = O^p(L_\beta)$.

Let $Z_\beta < N \leq Q_\beta$ with $N/Z_\beta = \bar{N}$ a minimal normal subgroup of G_β/Z_β. Assume that $|\bar{N}| < q_\beta^2$. From (5.7) we have $[N, O^p(L_\beta)] = 1$ and thus $N \leq Z(L_\beta)$. Hence, $Z(L_\beta)$ cannot be elementary abelian as $N \neq Z_\beta$. It follows that $Z(L_\beta) \leq Q_\alpha$ but $Z(L_\beta) \nleq D_\alpha$ by (10.4)(a). Since K normalizes $Z(L_\beta)$, we get that $Z(L_\beta)Q_{\alpha-1} \in Syl_p(L_{\alpha-1})$; in particular, $Q_\alpha = \langle Z(L_\beta)^{G_\alpha} \rangle D_\alpha$ and $D_\alpha \leq Z(Q_\alpha)$. This implies $R_\alpha \leq Z(L_\alpha) = 1$ by (10.2)(c). Thus $D_\alpha = Z_\alpha$, a contradiction.

We have proven that $|\bar{N}| \geq q_\beta^2$. Further, $|Q_\beta/N| < q_\beta^2$ implies as above that $Q_\beta = [Q_\beta, L_\beta] \leq N$. Thus we may assume that $|\bar{N}| = |Q_\beta/N| = q_\beta^2$, in particular, $q_\alpha = q_\beta$ and $N \cap Z_\alpha = Z_\beta$.

If $|N \cap Q_\alpha| > q_\beta^2$, it is easy to see with the help of (10.7) and the action of K on N that $Z_\alpha \leq N$. Thus, $N \nleq Q_\alpha$ and $|N \cap Q_\alpha| = q_\beta^2$. Let $\tilde{N} = N^x$, $x \in L_\alpha$ with $\alpha-1 = \beta^x$. Then $L_\alpha = \langle \tilde{N}, N \rangle Q_\alpha$, $Q_\alpha = (N \cap Q_\alpha)(\tilde{N} \cap Q_\alpha)D_\alpha$, and $[\tilde{N} \cap Q_\alpha, N \cap Q_\alpha] \leq N \cap D_\alpha \cap \tilde{N} = 1$. It follows that $\tilde{N} \cap Q_\alpha$ operates quadratically on N/Z_β and $W = (N \cap Q_\alpha) \times (\tilde{N} \cap Q_\alpha)$ is L_α-invariant. As both W/Z_α and Z_α are natural L_α/Q_α-modules we get from (5.17) that W is abelian. On the other hand, $|W \cap Q_{\alpha'}| = q_\beta^2$, and since W is transitive on $\Delta(\beta) \setminus \{\alpha\}$ we have $W \cap Q_{\alpha'} = W \cap G_\beta^{(2)}$. Hence $G_\beta^{(2)} \neq Z_\beta$ and $G_\beta^{(2)} \neq Q_\beta$. So $G_\beta^{(2)}$ satisfies the defining properties of N. But we have shown above that $N \nleq Q_\alpha$, this is a contradiction. □

<u>Remark</u>: By using the representation theory of the group $SL_2(q_\beta)$ it is not difficult to show that either $q_\alpha = q_\beta$ or $q_\alpha^3 = q_\beta$. In particular, the L_β/Q_β-module Q_β/Z_β can then be identified, from which it follows that L_β/Q_β is, in fact, isomorphic to $SL_2(q_\beta)$. One further interesting point is that although Q_β/Z_β is irreducible as a G_β/Q_β-module it is not in

general irreducible as an L_β/Q_β-module. Indeed, it is not irreducible precisely when $q_\alpha = q_\beta = 4$. By using this fact one can construct in this case a "subamalgam of G" which is parabolic isomorphic to J_2. We will say more on this in the last chapter.

11. The case $b = 1$ and $[Z_\alpha, Z_{\alpha'}] = 1$.

(11.0) Hypothesis. In this chapter Hypothesis B' holds and furthermore:

(a) $[Z_\alpha, Z_{\alpha'}] = 1$,

(b) $b = 1$.

We will show under this hypothesis that $r = s-1$ and stabilizers of paths of length s are p'-groups. In addition, we will give the chief factors of G_δ in Q_δ, $\delta \in \Gamma$. In a later chapter this will lead to the conclusion that G is locally isomorphic to X for $X \in \{PSp_4(q)$ (q odd), $U_4(q)$, $U_5(q)\}$.

Notation. $V_\beta = \langle (Z_\alpha \cap Q_\beta)^g \mid g \in G_\beta \rangle$.

We call the reader's attention to the fact that the definition of V_β in this chapter differs from the usual one since $Z_\alpha \not\leq Q_\beta$.

(11.1)(a) $\Omega_1(Z(L_\beta)) = Z_\beta$,

(b) $Z(L_\alpha) = 1$,

(c) $[Q_\beta, Z_\alpha, Z_\alpha] = 1$.

Proof: (a) and (b) follow as in (8.1), and (c) is an obvious consequence of $b = 1$.

(11.2)(a) L_β/Q_β <u>is neither a Ree group nor isomorphic to</u> $L_2(q_\beta)$,

(b) $Z_\alpha Q_\beta/Q_\beta = \Omega_1(Z(S/Q_\beta))$,

(c) $Z_\delta \cap Q_\beta \nleq Q_\alpha$, <u>for any</u> $\delta \in \Delta(\beta)\smallsetminus\{\alpha\}$.

<u>Proof</u>: By (11.1)(c) Z_α operates quadratically on $Q_\beta/\phi(Q_\beta)$, and similar on $\phi(Q_\beta)/\phi(\phi(Q_\beta))$, etc. If L_β/Q_β does not satisfy (a), (5.9)(a) implies that $[O^p(L_\beta),Q_\beta] = 1$. It follows that $C_{L_\beta}(Q_\beta) \nleq Q_\beta$, a contradiction.

Note that $Z_\alpha Q_\beta/Q_\beta$ is elementary abelian and K-invariant. Thus (b) follows from (a) and (5.9).

Assume that $Z_\delta \cap Q_\beta \leq Q_\alpha$, for some $\delta \in \Delta(\beta)\smallsetminus\{\alpha\}$, then $Z_\delta \cap Q_\beta$ is centralized by $L = \langle Z_\alpha, Z_\delta\rangle$ which is not a p-group. In addition, $[Q_\beta, Z_\delta] \leq Q_\beta \cap Z_\delta$ and $[Q_\beta, O^p(L), O^p(L)] = 1$. It follows that $O^p(L) \leq C_{L_\beta}(Q_\beta) \nleq Q_\beta$, a contradiction. □

<u>Notation</u>. Let $\tilde\gamma = (\alpha-2,\alpha-1,\alpha,\beta,\alpha+2)$ be a path of length 4, which is stabilized by K. By (11.2)(c) we have

$$Z_{\alpha+2} \cap Q_\beta \nleq Q_\alpha \quad\text{and}\quad Z_{\alpha-2} \cap Q_{\alpha-1} \nleq Q_\alpha.$$

We set $L = \langle Z_\alpha, Z_{\alpha+2}\rangle$ and $V = (Z_\alpha \cap Q_\beta)(Z_{\alpha+2} \cap Q_\beta)$.

(11.3) <u>Set</u> $D = Q_\alpha \cap Q_{\alpha+2}$, $\bar L = LQ_\beta/D$, <u>and</u> $A = (Z_{\alpha+2} \cap Q_\beta)Q_\alpha/Q_\alpha$. <u>Then the following hold</u>:

(a) $D \cap V = Z_\alpha \cap Z_{\alpha+2}$ <u>and</u> $|A| = |\overline{Z_{\alpha+2} \cap Q_\beta}|$,

(b) $\bar V = (\overline{Z_\alpha \cap Q_\beta}) \times (\overline{Z_{\alpha+2} \cap Q_\beta})$,

(c) $|A| = q_\alpha$, <u>or</u> L_α/Q_α <u>is a Ree group and</u> $|A| = q_\alpha^2$,

(d) $|A| \geq q_\beta$,

(e) $V' \neq 1$,

(f) $[Z_\alpha,A,A] \neq 1$,

(g) <u>if</u> $q_\beta^2 > |A|$, <u>then</u> q_β <u>is odd</u>, $|A| = q_\beta$, $L/L \cap Q_\beta \simeq SL_2(q_\beta)$, <u>and</u> $\bar{V}$ <u>is a natural</u> $L/L \cap Q_\beta$-<u>module</u>,

(h) $Q_\beta = V(Q_\alpha \cap Q_\beta)$.

<u>Proof</u>: It follows from (11.2)(a) and (b) and (5.2) - (5-4) that

$$L/L \cap Q_\beta \simeq SL_2(q_\beta) \quad \text{or} \quad Sz(q_\beta) \quad (\text{resp. } D_{10}).$$

In both cases there exists $x \in L$ so that $\alpha^x = \alpha+2$. So $Z_{\alpha+2} \cap Q_\alpha$ is normalized by L and we have $Z_{\alpha+2} \cap Q_\alpha = Z_\alpha \cap Z_{\alpha+2} = Z_\alpha \cap Q_{\alpha+2}$. Now the operation of K on $Z_{\alpha+2} \cap Q_\beta$ and again (5.2) - (5.5) yield (a),(b) and (c). In addition the quadratic action of Z_α on $\bar{V}$ together with (11.2) and (5.12) implies (d), namely

$$q_\beta \leq |A| .$$

Assume that $V' = 1$. Then A operates quadratically on Z_α, and again (5.12) together with (11.1)(b) imply that $L_\alpha/Q_\alpha \simeq SL_2(q_\alpha)$ and Z_α is a natural $SL_2(q_\alpha)$-module. But now $[Q_\beta,Z_\alpha] \leq Z_\alpha \cap Q_\beta = Z_\beta$ and $O^p(L_\beta) \leq C_{L_\beta}(Q_\beta)$, a contradiction. Hence, (e) holds, and this together with (5.8) implies (f).

Assume now that $q_\beta^2 > |A|$. Then again (5.12) yields that

$L/L \cap Q_\beta \simeq SL_2(q_\beta)$ and $\bar{V}$ is a natural $SL_2(q_\beta)$-module. In particular, $|A| = q_\beta$ and by (e) and (5.13) q_β is odd.

It remains to prove (h). Assume first that q_β is odd. As above $L/L \cap Q_\beta \simeq SL_2(q_\beta)$, and there exists an involution t_β in $K \cap L$ so that $[L,t_\beta] \leq Q_\beta$. From (b) and (5.9)(a) we get

$$[\bar{V},t_\beta] = \bar{V} .$$

On the other hand, we have $[Q_\beta,L] \leq V$ and so

$$\bar{Q}_\beta = \bar{V} \times \bar{C}$$

where $\bar{C} = C_{\bar{Q}_\beta}(\bar{t}_\beta)$. By definition of D we have $\overline{C \cap Q_\alpha} = 1$. It follows that $\overline{Q_\beta \cap Q_\alpha} = \overline{Z_\alpha \cap Q_\beta}$ and thus

$$\bar{Q}_\beta/\overline{Q_\beta \cap Q_\alpha} \simeq (\overline{Z_{\alpha+2} \cap Q_\beta}) \times \bar{C}$$

where the obvious isomorphism is compatible with the action of K. Hence, $Q_\beta Q_\alpha/Q_\alpha$ can be decomposed into the direct product of two K-invariant subgroups and (5.1) - (5.5) imply that either $\bar{C} = 1$ or $L_\alpha/Q_\alpha \simeq \mathrm{Ree}(q_\alpha)$. In the first case we are done. In the second case by (5.5) $|S/Q_\alpha Q_\beta| = q_\alpha$ and $|Q_\alpha Q_\beta/Q_\alpha| = q_\alpha^2$. But now $L_\beta/Q_\beta \simeq U_3(q_\beta)$ or $SU_3(q_\beta)$ and $S/Q_\alpha Q_\beta$ is a square, contradicting (5.5).

Assume now that q_β is even. Then L_α/Q_α is not a Ree group, and it follows from (c) that $A = \Omega_1(Z(S/Q_\alpha))$ and therefore $\Omega_1(Q_\beta/Q_\beta \cap Q_\alpha) \simeq \overline{Z_{\alpha+2} \cap Q_\beta}$. Thus, it suffices to show the existence of a direct decomposition

$$\bar{Q}_\beta/\overline{Q_\alpha \cap Q_\beta} \simeq \overline{Z_{\alpha+2} \cap Q_\beta} \times \bar{C}$$

to conclude that $\bar{C} = 1$. On the other hand, as seen above, such a decomposition clearly exists, if there is $y \in L$ which operates fixed-point-freely on $\bar{V}$. But such an element exists by (5.11). □

(11.4) $S = Q_\alpha Q_\beta$.

<u>Proof</u>: Assume that $S \neq Q_\alpha Q_\beta$. Then $L_\delta/Q_\delta \not\simeq L_2(q_\delta)$ or $SL(q_\delta)$ for every $\delta \in \Gamma$. Set $t = |S/Q_\alpha Q_\beta|$. We compute t in L_α and L_β. By (11.2), (5.3) and (5.4) we have one of the following cases:

(i) $t = q_\beta$ and $L_\beta/Q_\beta \simeq Sz(q_\beta)$,

(ii) $t = q_\beta^2$ and $L_\beta/Q_\beta \simeq U_3(q_\beta)$ or $SU_3(q_\beta)$,

(iii) $t = 2 = q_\beta$ and $L_\beta/Q_\beta \simeq U_3(2)$ or $SU_3(2)$.

In L_α we get with (5.3) - (5.5) that

(iv) $t = q_\alpha^2$ and $L_\alpha/Q_\alpha \simeq U_3(q_\alpha)$ or $SU_3(q_\alpha)$,

(v) $t \geq q_\alpha$ and L_α/Q_α is a Ree- or Suzuki group ,

(vi) $t = q_\alpha = 2$ and $L_\alpha/Q_\alpha \simeq U_3(2)$ or $SU_3(2)$.

If $q_\beta^2 \leq q_\alpha$, then only (ii) and (v) apply and $q_\beta^2 = q_\alpha$. Hence, q_α is a square which contradicts (5.3) and (5.5).

Assume now that $q_\beta^2 > q_\alpha$. Then (11.3)(c) and (g) apply. In particular, q_β is odd. So we are in the cases (ii) and (iv) or (v), and $t = q_\beta^2 = q_\alpha^2$. It follows (by the action of K) that $|Z_\alpha/Z_\alpha \cap Q_\beta| = q_\beta = |V_\beta/V_\beta \cap Q_\alpha| = |V_\beta/C_{V_\beta}(Z_\alpha)|$. But now (5.12) implies that $L_\beta/Q_\beta \simeq SL_2(q_\beta)$ contradicting our assumption. □

(11.5) $L_\alpha/Q_\alpha \simeq L_2(q_\alpha)$, $SL_2(q_\alpha)$ or D_{10} and $S = Q_\alpha(Z_{\alpha+2} \cap Q_\beta)$.

Proof: By (11.3)(h) and (11.4) $S = Q_\alpha V$. Hence, L_α/Q_α has elementary abelian Sylow p-subgroups, and by (11.3)(c) $S = Q_\alpha(Z_{\alpha+2} \cap Q_\beta)$.

□

(11.6) Set $Z = \bigcap_{\delta \in \Delta(\beta)} Q_\delta$. Then the following hold:

(a) $Z = Z_\beta$,

(b) $Z_\beta \cap Q_{\alpha-1} = 1$,

(c) $|Z_\beta| = q_\beta$ and $|Z_\alpha| = q_\beta^2 q_\alpha$,

(d) $q_\alpha = q_\beta$ or $q_\alpha = q_\beta^2$.

Proof: Set $Z_o = Z \cap Q_{\alpha-1}$ and $W = \langle Z_{\alpha-2} \cap Q_{\alpha-1}, Z_{\alpha+2} \cap Q_\beta \rangle$. By (11.5) $WQ_\alpha = L_\alpha$ and W is transitive on $\Delta(\alpha)$.

Note that Z centralizes $\langle Z_\alpha^{L_\beta} \rangle$ and so $C_{L_\beta}(Z)$ operates transitively on $\Delta(\beta)$. As $Z_o \cap Q_{\alpha-2}$ is centralized by W we conclude from (3.2)(c) that $Z_o \cap Q_{\alpha-2} = 1$; in particular, again by (11.5), Z_o is elementary abelian.

From (11.3)(a) (applied to $(\alpha,\alpha-2)$) we get $[Z_o, Z_{\alpha-2} \cap Q_{\alpha-1}] \leq Z_\alpha \cap Z_{\alpha-2}$, and so Z_oZ_α is normal in L_α. Hence, $[Z_oZ_\alpha, Q_\alpha]$ is normal, too. As $[Z_oZ_\alpha, Q_\alpha] = [Z_o, Q_\alpha] \leq Z_o$ we get $[Z_o, Q_\alpha] = 1$ by (3.2)(c), and Z_o is centralized by $Q_\alpha(Z_{\alpha+2} \cap Q_\beta) = S$ and so $Z_o \leq Z_\beta$.

Note that $Z_{\alpha+2} \cap Z_\alpha$ centralizes $Q_\alpha V = S$ and so $Z_{\alpha+2} \cap Z_\alpha = Z_\beta$, and with the same argument $Z_{\alpha-2} \cap Z_\alpha = Z_{\alpha-1}$. If $Z_\beta \leq Q_{\alpha-1}$ and $Z_{\alpha-1} \leq Q_\beta$, then $[Z_\beta, Z_{\alpha-2} \cap Q_{\alpha-1}] \leq Z_{\alpha-1}$ and $[Z_{\alpha-1}, Z_{\alpha+2} \cap Q_\beta] \leq Z_\beta$ and $Z_\beta Z_{\alpha-1}$ is normal in $WQ_\alpha = L_\alpha$. It follows that $Z_\alpha = Z_\beta Z_{\alpha-1}$, and $Z_{\alpha+2} \cap Q_\beta$ operates quadratically on Z_α, contradicting (11.3)(f).

Thus, we may assume without loss that $Z_\beta \not\leq Q_{\alpha-1}$. Assume further that $Z_o \neq 1$. Since $Z_o \leq Z_\beta$ we have $Z_\beta \cap Q_{\alpha-1} \neq 1$, and the action of K on Z_β together with (11.2)(a) imply that

$$(*) \qquad Z_\alpha = Z_\beta(Z_\alpha \cap Z_{\alpha-2}) = Z_\beta Z_{\alpha-1}.$$

Hence, (11.5) and (5.10)(c) yield

$$C_{Z_\alpha}(Z_{\alpha+2} \cap Q_\beta) = Z_\beta = C_{Z_\alpha}(t) \quad \text{for} \quad t \in (Z_{\alpha+2} \cap Q_\beta) \setminus Q_\alpha.$$

If q_α is even, then this equation implies that $Z_{\alpha+2} \cap Q_\beta$ operates quadratically on Z_α, again contradicting (11.3)(f). So q_α and q_β are odd. It follows as in (11.3) that $L/L \cap Q_\beta \simeq SL_2(q_\beta)$, and there exists an involution $t_\beta \in L \cap K$ so that $[L, t_\beta] \leq Q_\beta$. By (5.9)(a) t_β operates fixed-point-freely on V/Z_β.

Let $y \in N_{L_\alpha}(K)$ so that $\beta^y = \alpha-1$. Then t_β normalizes L^y, and since by (*) $[Z_\alpha, t_\beta] \leq Z_{\alpha-1}$, we get from (5.1) that $[L^y, t_\beta] \leq Z_{\alpha-1}$. In particular $[Z^y_{\alpha+2} \cap Q_{\alpha-1}, t_\beta] \leq Z_{\alpha-1} \leq Q_\alpha$, and again by (5.1) we have $[L_\alpha, t_\beta] \leq Q_\alpha$. But now $[Z_{\alpha+2} \cap Q_\beta, t_\beta] \leq Z_\beta$ and t_β does not operate fixed-point-freely on V/Z_β, a contradiction.

We have proven that $Z_o = 1$. Now $[Q_{\alpha-1}, Z, Z] \leq [Q_{\alpha-1} \cap Q_\alpha, Z] \leq Q_{\alpha-1} \cap Z = 1$ and so by (5.9) and the action of K (a) and (b) follow;

in particular $|Z_\beta| = q_\beta$. As a consequence of (11.3)(a) and (b) and (11.5) $|Z_\alpha| = q_\beta q_\alpha |Z_\alpha \cap Z_{\alpha+2}|$, and (a) implies $Z_\alpha \cap Z_{\alpha+2} = Z_\beta$, hence (c) follows.

From (11.5) and (5.7) we derive $q_\beta^2 \geq q_\alpha \geq q_\beta$. If $q_\beta^2 > q_\alpha$, then (11.3) implies $q_\alpha = q_\beta$, and (d) follows. □

(11.7) Assume that S/Q_β is elementary abelian. Then the following hold:

(a) $L_\beta/Q_\beta \simeq SL_2(q_\beta)$,

(b) $L_\alpha/Q_\alpha \simeq L_2(q_\alpha)$,

(c) $|S| = q_\alpha^2 q_\beta^2$,

(d) $q_\alpha = q_\beta \equiv 1(2)$ or $q_\alpha = q_\beta^2$,

(e) $Q_\alpha = Z_\alpha$ and $Q_\beta = V_\beta$,

(f) $4 = r = s-1$, and the stabilizers of paths of length s are p'-groups.

Proof: By (11.2) and (11.4) we have $S = Z_\alpha Q_\beta$; in particular by (11.2)(a) $L_\beta/Q_\beta \simeq SL_2(q_\beta)$ or D_{10}. It follows that $Z = Q_\alpha \cap Q_{\alpha+2}$ is normal in L_β, and (11.3) and (11.6) imply (c), (d) and (e).

If $L_\beta/Q_\beta \simeq D_{10}$, then $q_\beta = 2$ and V is extra special of order 2^5. Since there are elementary abelian subgroups of order 2^3 in V (for example $Z_\alpha \cap Q_\beta$), we have $5 \nmid |\mathrm{Aut}\ (V)|$. This contradiction shows (a).

The order of S implies that $r \leq 4$ and that (f) holds, if $r = 4$. Note that by (a) and (b) a Sylow p-subgroup of $G_{\alpha+2} \cap G_\beta$ (resp. $G_{\alpha-1} \cap G_\alpha$) operates transitively on $\Delta(\alpha+2) \setminus \{\beta\}$ (resp. $\Delta(\alpha+2) \setminus \{\alpha\}$). Hence, in view of (3.4)(c) is suffices to show that

$(Q_\alpha \cap Q_{\alpha+2})Q_{\alpha-1} \in Syl_2(L_{\alpha-1})$ and $(Q_{\alpha-1} \cap Q_\beta)Q_{\alpha+2} \in Syl_2(L_{\alpha+2})$.

The first claim follows from (11.6)(b) and the action of K. Set $E = Z_\alpha \cap Q_\beta \cap Q_{\alpha-1}$. By (11.6) we have $E \cap Q_{\alpha+2} = 1$ and $|E| = q_\alpha$. Now the second claim follows.

It remains to prove (b). From (11.5) and (d) we get $L_\alpha/Q_\alpha \simeq L_2(q_\alpha)$ or $SL_2(q_\alpha)$. Thus, we may assume that $L_\alpha/Q_\alpha \simeq SL_2(q_\alpha)$ and q_α is odd. Set $<t_\delta> = K \cap L_\delta \cap G_\delta^{(1)}$ for $\delta \in \tilde{\gamma}$. By (11.6)(b) and (5.1) we have $[L_{\alpha-1}, t_\beta] \leq Q_{\alpha-1}$. On the other hand, (6.9) yields $t_\beta = t_{\alpha-2}$ and so also $[L_{\alpha-2}, t_\beta] \leq Q_{\alpha-2}$. Hence, (3.2)(c) applied to $C_G(t_\beta)$ and $(\alpha-2, \alpha-1)$ implies a contradiction. □

Remark: It is easy to see that $[Z_\alpha, S, S] \neq 1 = [Z_\alpha, S, S, S]$. Using knowledge of the irreducible $SL_2(q_\alpha)$-modules it can be shown in the case $q_\alpha = q_\beta$ that Z_α is isomorphic to the module of homogenous polynomials of degree 2 in two indeterminates. In the case $q_\alpha = q_\beta^2$, Z_α is the "orthogonal module" for $L_2(q_\alpha) \simeq O_4(q_\alpha^{1/2})$. We will not need this detailed information in the subsequent analysis.

(11.8) Assume that S/Q_β is not elementary abelian. Then the following hold:

(a) $q_\alpha = q_\beta^2$,

(b) $L_\alpha/Q_\alpha \simeq SL_2(q_\alpha)$,

(c) $L_\beta/Q_\beta \simeq U_3(q_\beta)$ or $SU_3(q_\beta)$,

(d) $|S| = q_\alpha^4 q_\beta^2$,

(e) Q_α/Z_α is a natural L_α/Q_α-module ,

(f) Q_β <u>is special of order</u> q_β^7 <u>with center</u> Z_β,

(g) $4 = r = s-1$, <u>and stabilizers of paths of length</u> s <u>are</u> p'-<u>groups</u>.

<u>Proof</u>: By (11.2)(a) we have $L_\beta/Q_\beta \simeq U_3(q_\beta)$, $SU_3(q_\beta)$ or $Sz(q_\beta)$. Set $Z = \bigcap_{\delta\in\Delta(\beta)} Q_\beta$. From (5.3) and (5.4) we conclude that either $Z = Q_\alpha \cap Q_{\alpha+2}$ and $L_\beta/Q_\beta \simeq Sz(q_\beta)$ or $Z = Q_\alpha \cap Q_{\alpha+2} \cap Q_\lambda$ for some $\lambda \in \Delta(\beta) \setminus \{\alpha,\alpha+2\}$. Together with (11.6) we get

(i) $|Q_\beta| = q_\alpha^2 q_\beta$ and $L_\beta/Q_\beta \simeq Sz(q_\beta)$,

or

(ii) $|Q_\beta| \leq q_\alpha^3 q_\beta$ and $L_\beta/Q_\beta \simeq U_3(q_\beta)$ or $SU_3(q_\beta)$.

Assume first that (i) holds. Then $Q_\beta = V$ and $[Q_\alpha,V] \leq V \cap Q_\alpha \leq Z_\alpha$ by (11.3), i.e. Q_α operates quadratically on V. Now (5.8) implies that $\phi(Q_\alpha) \leq Q_\beta$, and by (11.4) S/Q_β is elementary abelian, a contradiction.

Assume now that (ii) holds. Then (5.7) and (11.6)(d) yield (a), (c) and (d), and (11.3)(e) proves (f). In addition, $|Q_\alpha/Z_\alpha| = q_\alpha^2$ by (11.6), and $Z_{\alpha+2} \cap Q_\beta$ operates quadratically on Q_α/Z_α. Now (5.12) and (11.5) imply (b) and (e).

It remains to prove (g). We argue as in (11.7). Note that $|Q_\beta \cap Q_{\alpha-1}| \geq q_\alpha$ and $|Q_\alpha \cap Q_{\alpha+2}| \geq q_\beta^3$. Hence, it suffices to show that $Q_\beta \cap Q_{\alpha-1} \cap Q_{\alpha+2} = 1$ and $Q_\alpha \cap Q_{\alpha+2} \cap Q_{\alpha-1} = 1$.

Set $Y = Q_{\alpha-1} \cap Q_\beta \cap Q_{\alpha+2}$ and note that $Q_\alpha \cap Q_{\alpha+2} \cap Q_{\alpha-1} \leq Y$. We have $[Y,Z_{\alpha-2} \cap Q_{\alpha-1}] \leq Z_{\alpha-1} \leq Z_\alpha$. Hence, YZ_α/Z_α is central in Q_α/Z_α

and (e) implies $Y \leq Z_\alpha$. On the other hand, $Q_{\alpha+2} \cap Z_\alpha = Z_\alpha \cap Z_{\alpha+2}$ and by (11.6) $Z_\alpha \cap Z_{\alpha+2} = Z_\beta$. It follows that $Y \leq Z_\beta \cap Q_{\alpha-1}$, and again (11.6) implies $Y = 1$. □

Remark: As we commented above after (11.7) it is possible to show using knowledge of the irreducible $SL_2(q_\alpha)$-modules that Z_α is the "orthogonal module" for $SL_2(q_\alpha) \simeq O_4(q_\alpha^{1/2})$. Furthermore, the module Q_β/Z_β carries the structure of the natural 3-dimensional $GF(q_\beta^2)$-module for $L_\beta/Q_\beta \simeq SU_3(q_\beta)$.

12. The case $[V_{\alpha'}, V_\beta \cap Q_{\alpha'}] \neq 1$.

In this chapter we continue the analysis of the structure of G_α and G_β under the following hypothesis.

(12.0) Hypothesis. In this chapter Hypothesis (8.0) holds.

We will show that G is of type ${}^2F_4(2)'$ or $r = s-1$ and stabilizers of path of length s are p'-groups. In a later chapter this second case will lead to the conclusion that G is locally isomorphic to ${}^2F_4(q)$ or of type ${}^2F_4(2)$.

We first sum up the results of chapter 8.

(12.1)(a) $b = 3$.

(b) $q_\alpha = q_\beta$.

(c) $Q_\alpha Q_\beta = S$.

(d) $Z(L_\alpha) = 1$.

(e) $L_\alpha/Q_\alpha \simeq SL_2(q_\alpha)$, and Z_α is a natural L_α/Q_α-module.

(f) $[V_\beta, Q_\beta] = Z_\beta$.

(g) $[V_\beta \cap Q_{\alpha'}, V_{\alpha'} \cap Q_\beta] = 1$.

(h) L_β/Q_β is not a Ree group.

Proof: This is (8.1), (8.4), (8.6), and (8.7). □

In the following we will use (12.1)(a) - (d) without reference.

(12.2) L_β/Q_β <u>is not isomorphic to</u> $U_3(2)$ <u>or</u> $SU_3(2)$.

<u>Proof</u>: Obviously, it does not matter whether we prove the assertion for L_β or $L_{\alpha'}$.

Assume first that $L_\beta/Q_\beta \simeq SU_3(2)$. Then $|K_\beta| = 3$ and $K_\beta \le G_\beta^{(1)}$. On the other hand, by (12.1)(e) $L_\alpha/Q_\alpha \simeq SL_2(2)$ and so $K_\beta \le G_\alpha^{(1)}$, contradicting (3.2)(c).

Assume now that $L_{\alpha'}/Q_{\alpha'} \simeq U_3(2)$ and set $F = \langle Z_\alpha^{L_{\alpha'}}\rangle Q_{\alpha'}/Q_{\alpha'}$. Then F is a Frobenius group of order 18 with elementary abelian kernel. In addition $|V_{\alpha'}/C_{V_{\alpha'}}(Z_\alpha)| = |V_{\alpha'}/V_{\alpha'} \cap Q_\alpha| = 4$ and we get $|[V_{\alpha'}/Z_{\alpha'},F]| = 2^4$, contradicting (5.7). □

(12.3) <u>Set</u> $H = V_\beta \cap V_{\alpha'}$. <u>Then the following hold</u>:

(a) $|V_\beta/V_\beta \cap Q_{\alpha'}| = |V_\beta \cap Q_{\alpha'}/V_\beta \cap Q_\lambda| = q_{\alpha'}$ <u>for some</u> $\lambda \in \Delta(\alpha') \smallsetminus \{\alpha+2\}$,

(b) H <u>is normal in</u> $L_{\alpha+2}$,

(c) $[H,Q_{\alpha+2}] = Z_{\alpha+2}$,

(d) $[V_{\alpha'},L_{\alpha'}] = V_{\alpha'}$.

<u>Proof</u>: Let $\tilde{K}$ be a complement to a Sylow p-subgroup in $G_\beta \cap G_{\alpha+2}$ and choose $\lambda \in \Delta(\alpha')$ so that $[V_\beta \cap Q_{\alpha'},Z_\lambda] \neq 1$. By (12.1)(e) we may choose $\tilde{K} \le G_{\alpha'}$. Hence, $\tilde{K}$ normalizes V_β and $V_\beta \cap Q_{\alpha'}$.

If $L_{\alpha'}/Q_{\alpha'} \simeq D_{10}$, then $\tilde{K} \le G_{\alpha'}^{(1)}$ and so $\tilde{K} \le G_\lambda$. If $L_{\alpha'}/Q_{\alpha'} \not\simeq D_{10}$, then $Q_{\alpha+2}$ is transitive on $\Delta(\alpha') \smallsetminus \{\alpha+2\}$, and there exists a $Q_{\alpha+2}$-conjugate $\tilde{\tilde{K}}$ of $\tilde{K}$ in G_λ. In both cases $V_\beta \cap Q_{\alpha'}$ is normalized by $\tilde{K}$ resp. $\tilde{\tilde{K}}$ and (12.1)(b) and (e) yield $|V_\beta \cap Q_{\alpha'}/V_\beta \cap Q_\lambda| = q_\lambda = q_{\alpha'}$. Since $\tilde{K}$ normalizes V_β, we get (a) from (12.1)(h) and (5.9)(b).

Note that $\langle Q_\beta, Q_{\alpha'} \rangle Q_{\alpha+2} = L_{\alpha+2}$ and $Q_{\alpha+2} = (Q_\beta \cap Q_{\alpha+2})(Q_{\alpha'} \cap Q_{\alpha+2})$. Hence, by (12.1)(f) H is normal in $L_{\alpha+2}$ and $[H,Q_{\alpha+2}] \leq Z_{\alpha+2}$. Since $[H,Q_{\alpha+2}]$ is normal in $L_{\alpha+2}$, we get by (12.1)(e) $[H,Q_{\alpha+2}] = Z_{\alpha+2}$ or $H \leq Z(Q_{\alpha+2})$.

In the first case $Z_{\alpha+2} \leq [V_{\alpha'}, L_{\alpha'}]$ and (d) follows. Assume that $H \leq Z(Q_{\alpha+2})$. After conjugation in $L_{\alpha'}$ we get from (a):

$$|V_\sigma \cap Q_{\alpha'}/V_\sigma \cap Q_{\alpha+2}| = q_{\alpha+2} \quad \text{for some } \sigma \in \Gamma \text{ with } d(\sigma,\alpha') = 2.$$

On the other hand, $|H/H \cap Q_\sigma| \leq q_\sigma$ and by (12.1)(g) (applied to (α',σ)) $[H \cap Q_\sigma, V_\sigma \cap Q_{\alpha'}] = 1$. It follows that

$$q_{\alpha+2} \geq |H/C_H(V_\sigma \cap Q_{\alpha'})|$$

and we conclude from (12.1)(e) and $Z(L_{\alpha+2}) = 1$ that $H = Z_{\alpha+2}$.

Now set $\bar{V}_{\alpha'} = V_{\alpha'}/Z_{\alpha'}$ and $L = \langle V_\beta^{L_{\alpha'}} \rangle Q_{\alpha'}/Q_{\alpha'}$. By (12.1)(f) $\bar{V}_{\alpha'}$ is an L-module, and L is generated by two or by three conjugates of $V_\beta Q_{\alpha'}/Q_{\alpha'}$. In addition, we have $[\bar{V}_{\alpha'}, V_\beta] = \bar{Z}_{\alpha+2}$ and $|\bar{Z}_{\alpha+2}| = q_{\alpha'}$. Hence, we get $|[\bar{V}_{\alpha'}, L]| \leq q_{\alpha'}^3$ and by (5.7) and (5.9) $L_{\alpha'}/Q_{\alpha'} \simeq SL_2(q_{\alpha'})$ and $|\bar{V}_{\alpha'}| = q_{\alpha'}^2$, and $\bar{V}_{\alpha'}$ is a natural $SL_2(q_{\alpha'})$-module. But this contradicts (8.5). □

(12.4)(a) $L_{\alpha'}/Q_{\alpha'} \simeq U_3(q_{\alpha'})$ <u>or</u> $SU_3(q_{\alpha'})$, $q_{\alpha'} > 2$, <u>or</u> $Sz(q_{\alpha'})$.

(b) $s \geq 4$.

<u>Proof</u>: Assume first that (a) does not hold. In view of (12.1)(h) and (12.2) we then have $L_{\alpha'}/Q_{\alpha'} \simeq D_{10}$, $L_2(q_{\alpha'})$ or $SL_2(q_{\alpha'})$, and (12.3)(a) implies $V_\beta Q_{\alpha'} \in Syl_p(L_{\alpha'})$; in particular $Q_\beta \cap Q_{\alpha+2} \leq V_\beta Q_{\alpha'}$.

Assume now that $Q_\beta \cap Q_{\alpha+2} \leq V_\beta Q_{\alpha'}$. Then conjugation in $L_{\alpha+2}$ gives $Q_{\alpha'} \cap Q_{\alpha+2} \leq V_{\alpha'} Q_\beta$. Since by (12.1)(f) $[Q_{\alpha'} \cap Q_\beta, V_\beta] \leq Z_\beta \leq V_{\alpha'}$, we get

$$[Q_{\alpha'} \cap Q_{\alpha+2}, V_\beta] \leq V_{\alpha'}.$$

Hence, V_β centralizes $Q_{\alpha'} \cap Q_{\alpha+2}/V_{\alpha'}$ which is of index $q_{\alpha'}$ in $Q_{\alpha'}/V_{\alpha'}$. Now (12.3)(a) and (5.12) imply either $[Q_{\alpha'}, O^p(L_{\alpha'})] \leq V_{\alpha'}$ or $L_{\alpha'}/Q_{\alpha'} \simeq SL_2(q_{\alpha'})$ and $V_\beta Q_{\alpha'} \in Syl_p(L_{\alpha'})$.

In the first case $Q_{\alpha'} \cap Q_{\alpha+2}$ is normal in $L_{\alpha'}$, contradicting (6.5). In the second case $Q_{\alpha+2} \leq V_\beta Q_{\alpha'} \cap V_{\alpha'} Q_\beta$ and $Q_{\alpha+2} = V_{\alpha'} V_\beta (Q_{\alpha'} \cap Q_\beta)$. By (12.1)(f) $[Q_{\alpha'} \cap Q_\beta, V_{\alpha'} \cap V_\beta] \leq Z_\beta \cap Z_{\alpha'} = 1$ and so $V_{\alpha'} \cap V_\beta$ is central in $Q_{\alpha+2}$, contradicting (12.3)(c).

We have shown that (a) holds and that $Q_\beta \cap Q_{\alpha+2} \not\leq V_\beta Q_{\alpha'}$. Now the action of $\tilde{K}$ (chosen as in (12.2)) on $Q_\beta \cap Q_{\alpha+2}$ yields

$$(*) \qquad (Q_\beta \cap Q_{\alpha'}) Q_{\alpha'} \in Syl_p(L_{\alpha'}).$$

It remains to prove (b). By (*) $Q_\beta \cap Q_{\alpha+2}$ is transitive on $\Delta(\alpha') \setminus \{\alpha+2\}$ and by (12.1)(e) and (12.3)(a) $V_\beta \cap Q_{\alpha'}$ is transitive on $\Delta(\lambda) \setminus \{\alpha'\}$ for $\lambda \in \Delta(\alpha')$. This implies (b). □

<u>Notation</u>. In view of (12.4)(b) and (6.7) we may assume that

$K \leq G_\gamma$. In addition, we set $q = q_\alpha = q_\beta$.

(12.5) $L_{\alpha'}/Q_{\alpha'} \simeq Sz(q)$.

Proof: Note that by (12.1)(e) $|Z_{\alpha'}| = q$. According to (12.4)(a) we may assume that $L_{\alpha'}/Q_{\alpha'} \simeq U_3(q)$ or $SU_3(q)$, and by (5.7) we get $|V_{\alpha'}| \geq q^7$. Now (12.4)(b), (3.4)(c) and (12.3)(b) yield for any path $(\delta_0,\dots,\delta_3)$ of length 3 with $\delta_0 \in \beta^G$:

$$(*) \qquad (V_{\delta_0} \cap Q_{\delta_2})Q_{\delta_3} \in Syl_p(L_{\delta_3}).$$

Pick $\alpha-2 \in \Gamma$ with $d(\alpha-2,\beta) = 3 = d(\alpha-2,\alpha)+1$ and $\alpha'+3 \in \Gamma$ with $d(\alpha'+3,\alpha') = 3 = d(\alpha'+3,\alpha+2) - 1$. By (12.4)(b) we have $Z_{\alpha-2} \leq G_\beta$ but $Z_{\alpha-2} \not\leq G_{\alpha+2}$ and $Z_{\alpha'+3} \leq G_{\alpha'}$ but $Z_{\alpha'+3} \not\leq G_{\alpha+2}$. In addition, we have $|V_{\alpha'}/V_{\alpha'} \cap Q_{\alpha-2}| \leq q^4$ and $|V_{\alpha'}/V_{\alpha'} \cap Q_{\alpha'+3}| \leq q^2$, and from $|V_{\alpha'}| \geq q^7$ we conclude that $V_{\alpha'} \cap Q_{\alpha-2} \cap Q_{\alpha'+3} \neq 1$.

Pick $1 \neq x \in V_{\alpha'} \cap Q_{\alpha-2} \cap Q_{\alpha'+3}$. As $[Q_{\alpha'},x] \leq Z_{\alpha'}$ by (12.1)(f) we have that $|Q_{\alpha'}/C_{Q_{\alpha'}}(x)| \leq q$. On the other hand, $(Q_{\alpha+2} \cap Q_{\alpha'})Q_\beta \in Syl_p(L_\beta)$ since $s > 3$ and so $(C_{Q_{\alpha'}}(x) \cap Q_{\alpha+2})Q_\beta$ is of index at most q in a Sylow p-subgroup of $G_\beta \cap G_{\alpha+2}$. So $L = \langle C_{Q_{\alpha'}}(x) \cap Q_{\alpha+2}, Z_{\alpha-2}\rangle$ is transitive on $\Delta(\beta)$, and as $[L,x] = 1 = [Z_{\alpha+2},x]$ we have $\langle Z_{\alpha+2}^L\rangle = V_\beta$ centralizes x.

Set $A = \langle Z_\alpha, Z_{\alpha'+3}\rangle$, then A centralizes x, too. Choose $a \in A \setminus G_{\alpha+2}$ and $b \in L \setminus G_{\alpha+2}$. Then V_β^a and $V_{\alpha'}^b$ centralize x. But by $(*)$, $\langle V_{\alpha'}^b \cap Q_\beta, V_\beta^a \cap Q_{\alpha'}\rangle$ is transitive on $\Delta(\alpha+2)$ and an application of (3.2)(c) to $C_G(x)$ shows that $x = 1$, a contradiction. □

(12.6)(a) $|V_{\alpha'}| = q^5$ <u>and</u> $|Q_{\alpha'}/G^{(2)}_{\alpha'}| = q^4$.

(b) $V_{\alpha'}/Z_{\alpha'}$ <u>and</u> $Q_{\alpha'}/G^{(2)}_{\alpha'}$ <u>are irreducible</u> $L_{\alpha'}/Q_{\alpha'}$<u>-modules</u>.

(c) $|V_{\alpha'} \cap V_\beta| = q^3$.

<u>Proof</u>: Set $\tilde{L}_{\alpha'} = Z_\alpha O^2(L_{\alpha'})Q_{\alpha'}$ and $Z = C_{V_{\alpha'}}(\tilde{L}_{\alpha'})$. By (12.5) and the operation of K we have $\tilde{L}_{\alpha'} = \langle Z_\alpha, Z_\alpha^g\rangle Q_{\alpha'}$ for $g \in L_{\alpha'} \setminus G_{\alpha+2}$ and $\tilde{L}_{\alpha'} = L_{\alpha'}$ or $L_{\alpha'}/Q_{\alpha'} \simeq Sz(2)$. Note that $|V_{\alpha'}/V_{\alpha'} \cap Q_\alpha| \le q^2$ and note further that $[(Q_{\alpha'} \cap Q_\beta)V_{\alpha'}, Z_\alpha] \le V_{\alpha'} \le G^{(2)}_{\alpha'}$ and $|Q_{\alpha'}/(Q_{\alpha'} \cap Q_\beta)V_{\alpha'}| \le q^2$ by (12.1)(e) (applied to $\alpha+2$) and (12.3)(a) (applied to α'). It follows that $|Q_{\alpha'}/G^{(2)}_{\alpha'}| \le q^4$ and $|V_{\alpha'}/C_{V_{\alpha'}}(\langle Z_\alpha, Z_\alpha^g\rangle)| \le q^4$.

Since $Q_{\alpha'} \cap Q_{\alpha+2}$ is not normal in $L_{\alpha'}$ by (6.5), we conclude from (5.7) that $Q_{\alpha'}/G^{(2)}_{\alpha'}$ is an irreducible $L_{\alpha'}/Q_{\alpha'}$-module of order q^4. Set $Q = C_{V_{\alpha'}}(\langle Z_\alpha, Z_\alpha^g\rangle)$. Then Q is elementary abelian and $[Q, G^{(2)}_{\alpha'}] = 1$. Now the Three-subgroup Lemma applied to $[Q_{\alpha'}, \langle Z_\alpha, Z_\alpha^g\rangle, Q]$ and (12.1)(f) yield $Q = Z$. As above $V_{\alpha'}/Z$ is an irreducible $L_{\alpha'}/Q_{\alpha'}$-module of order q^4. It remains to prove $Z = Z_{\alpha'}$. In a counter example we have $L_{\alpha'}/Q_{\alpha'} \simeq Sz(2)$, $|Z_{\alpha'}| = 2 = |Z/Z_{\alpha'}|$ and $[Q_\beta \cap Q_{\alpha+2}, Z] = Z_{\alpha'}$, since by (12.4)(b) $(Q_\beta \cap Q_{\alpha+2})Q_{\alpha'} \in Syl_2(L_{\alpha'})$.

Since $V_\beta \le \tilde{L}_{\alpha'}$ we get $Z \le G^{(2)}_\beta$. Pick $\alpha-1 \in \Delta(\alpha) \setminus \{\beta\}$. Then, again by (12.4)(b) and (12.3)(a), $\langle V_{\alpha-1}, Q_{\alpha+2}\rangle Q_\beta = L_\beta$ and $Q_\beta = (V_{\alpha-1} \cap Q_\beta)(Q_\beta \cap Q_{\alpha+2})$. On the other hand, $Z \le V_\beta Q_{\alpha-1}$ and $[Z, V_{\alpha-1}] \le [V_\beta, V_{\alpha-1}] \le V_\beta$. It follows that ZV_β and thus $[ZV_\beta, Q_\beta]$ is normal in L_β; and since $Z_\beta Z_{\alpha'} = Z_{\alpha+2} \le [ZV_\beta, Q_\beta]$, we get $[ZV_\beta, Q_\beta] = V_\beta$. But $|Q_\beta/Q_\beta \cap \tilde{L}_{\alpha'}| \le 4$, $|Z/Z \cap V_\beta| = 2$, $[V_\beta, Q_\beta] = Z_\beta$, and so we have $|[ZV_\beta, Q_\beta]| = |V_\beta| \le 8$, a contradiction. □

(12.7) <u>The stabilizers of paths of length</u> 9 <u>are</u> <u>p'-groups</u>.

<u>Proof</u>: Let $\tilde{\gamma} = (\delta_0,\ldots,\delta_9)$ be a path of length 9. We arrange notation so that $\delta_0 \in \alpha^G$ and $\delta_9 \in \beta^G$. Since $s \geq 4$ we may assume that $(\delta_2,\ldots,\delta_5) = (\alpha,\ldots,\alpha')$. In addition, we have according to (12.3)(a), for any $\delta,\mu \in \beta^G$ and $\lambda \in \Delta(\mu)$ with $d(\delta,\mu) = 2$ and $d(\delta,\lambda) = 3$:

$$(*) \qquad |V_\delta/V_\delta \cap Q_\mu| = |V_\delta \cap Q_\mu/V_\delta \cap Q_\lambda| = q.$$

The operation of K on $Z_\alpha/Z_\alpha \cap Q_{\alpha'}$ yields

$$(**) \qquad |Z_\alpha/Z_\alpha \cap Q_{\alpha'}| = q.$$

Assume now that there exists a p-element $x \neq 1$ in $G_{\tilde{\gamma}}$. Note that $x \in Q_{\delta_i}$ and $[Z_{\delta_i},x] = 1$ for $i = 1,\ldots,8$. We define $A = C_{V_{\delta_1}}(x)$, $L = \langle Z_\alpha, Z_{\delta_8}\rangle$, $\tilde{L} = \langle A, V_{\alpha'}\rangle$ and $D = A \cap Q_{\alpha'}$. Then x centralizes A and L.

By (12.5) and (**) L is transitive on $\Delta(\alpha')$; in particular x centralizes $\langle Z^L_{\alpha+2}\rangle = V_{\alpha'}$ and thus $\tilde{L}$.

If $A \nleq Q_\beta$, then, as above, $\tilde{L}$ is transitive on $\Delta(\beta)$ and V_β centralizes x. Hence, according to (*) there exists $a \in L$ and $b \in \tilde{L}$ so that $\langle V^a_\beta \cap Q_{\alpha'}, V^b_{\alpha'} \cap Q_\beta\rangle$ is transitive on $\Delta(\alpha+2)$, and an application of (3.2)(c) to $C_G(x)$ yields a contradiction.

Assume now that $A \leq Q_\beta$. Since $[V_{\delta_1},x] \leq Z_{\delta_1}$ we have $|V_{\delta_1}/A| \leq q$ and so $A = V_{\delta_1} \cap Q_\beta$; in particular by (*) $|V_{\delta_1} \cap Q_\beta/V_{\delta_1} \cap Q_{\alpha+2}| = q$. If $D \leq Q_{\delta_6}$, then D is centralized by

$\langle V_{\delta_1}, Z_{\delta_6}\rangle$ which is transitive on $\Delta(\beta)$. On the other hand, by (12.6)(c) and (*), $V_{\delta_1} \cap V_\beta = V_{\delta_1} \cap Q_{\alpha+2}$ and $V_{\delta_1} \cap V_\beta \leq A$ whence $|V_{\delta_1} \cap V_\beta / D \cap V_\beta| \leq q$ and, by (12.6)(c), $|D \cap V_\beta| \geq q^2$. It follows that $(D \cap V_\beta)Z_\beta$ is normal in $O^2(L_\beta)Q_\beta$ but $D \cap V_\beta \not\leq Z_\beta$, contradicting (12.6)(a) and (b).

Thus, we have $D \not\leq Q_{\delta_6}$ and there exists $a \in L$ so that $D^a \not\leq Q_{\alpha+2}$, and $\langle A, D^a\rangle$ is transitive on $\Delta(\alpha+2)$. Now again (3.2)(c) applied to $C_G(x)$ yields a contradiction. □

(12.8) $s \geq 7$.

Proof: Note that by (12.1)(e) and (12.5) $|\Delta(\alpha) \setminus \{\beta\}| = q$ and $|\Delta(\beta) \setminus \{\alpha\}| = q^2$. Hence, there exist paths

$$\gamma_1 = (\alpha, \beta, \ldots, \alpha+6) \quad \text{and} \quad \gamma_2 = (\alpha-5, \ldots, \alpha, \beta)$$

of length 6 which are stabilized by K. Since $s \geq 4$ we get from (12.3)(a) that $V_{\alpha+3} \cap Q_\beta$ is transitive on $\Delta(\alpha) \setminus \{\beta\}$ and $V_{\alpha+3} \cap Q_{\alpha+5}$ is transitive on $\Delta(\alpha+6) \setminus \{\alpha+5\}$. Set $V = V_{\alpha+3} \cap Q_\beta \cap Q_{\alpha+5}$, and assume that $V \leq Q_\alpha$. Then $V = C_{V_{\alpha+3}}(Z_\alpha) = V_{\alpha+3} \cap V_\beta$ by (12.6), and from $V_{\alpha+3} = (V_\beta \cap V_{\alpha+3})(V_{\alpha+3} \cap V_{\alpha+5})$ we conclude that $V_{\alpha+3} \leq Q_{\alpha+5}$, contradicting $s \geq 4$. Hence $V \not\leq Q_\alpha$, and the action of K on V implies that V is transitive on $\Delta(\alpha) \setminus \{\beta\}$. A symmetric argument at $\alpha+6$ now shows that γ_1 is regular.

Set $D = Q_{\alpha-4} \cap Q_{\alpha-2} \cap Q_\alpha$. By (12.1)(e), (12.5) and (12.6) $|Q_{\alpha-2}| \geq q^{10}$ and $|Q_{\alpha-2}/D| \leq q^6$; i.e. $|D| \geq q^4$. Assume that D is not transitive on $\Delta(\beta) \setminus \{\alpha\}$. Then the operation of K yields $|D/D \cap Q_\beta| \leq q$. It follows that $|D \cap Q_{\alpha+2}| \geq q^2$, and we conclude from (12.7) that $D \cap Q_{\alpha+2} \cap Q_{\alpha-5} = 1$ and $D \cap Q_{\alpha+2}$ is transitive on $\Delta(\alpha-5) \setminus \{\alpha-4\}$. Thus, we have shown that G_{γ_2} is transitive on $\Delta(\alpha-5) \setminus \{\alpha-4\}$ or $\Delta(\beta) \setminus \{\alpha\}$. We fix notation so that G_{γ_2} is transitive on $\Delta(\alpha-5) \setminus \{\alpha-4\}$.

Now assume that $s < 7$, and let $\tilde{\gamma} = (\delta_0, \ldots, \delta_s)$ be a path of length s so that $G_{\tilde{\gamma}}$ is not transitive on $\Delta(\delta_0) \setminus \{\delta_1\}$. By (3.4)(c) we may assume that $\tilde{\gamma}$ is a subpath of γ_1 or γ_2 and $\delta_0 = \alpha$ or $\delta_0 = \alpha-5$, respectively. But now $G_{\gamma_i} \subseteq G_{\tilde{\gamma}}$, $i = 1$ or 2, and G_{γ_i} is transitive on $\Delta(\delta_0) \setminus \{\delta_1\}$, a contradiction. □

(12.9) One of the following holds:

(a) G is of type ${}^2F_4(2)'$ and $s = 7$, $r = 8$.

(b) G is of type ${}^2F_4(2)$ and $r = s-1 = 8$.

(c) $r = s-1 = 8$, $q > 2$, and

(c1) $L_\alpha/Q_\alpha \simeq SL_2(q)$, $L_\beta/Q_\beta \simeq Sz(q)$,

(c2) $Q_\beta/G_\beta^{(2)}$ and V_β/Z_β are irreducible $Sz(q)$-modules of order q^4; and Z_β and $G_\beta^{(2)}/V_\beta$ are central $Sz(q)$-modules of order q,

(c3) $G_\alpha^{(2)} = \langle (V_\beta \cap Q_{\alpha-1})^{G_\alpha} \rangle$ and $G_\alpha^{(3)} = V_\beta \cap V_{\alpha-1}$, $\alpha-1 \in \Delta(\alpha) \setminus \{\beta\}$,

(c4) $Q_\alpha/V_\alpha^{(2)}$, $G_\alpha^{(2)}/G_\alpha^{(3)}$ and Z_α are natural $SL_2(q)$-modules; $V_\alpha^{(2)}/G_\alpha^{(2)}$ is an irreducible $SL_2(q)$-module of order q^4; and $G_\alpha^{(3)}/Z_\alpha$ is a central $SL_2(q)$-module of order q.

(c5) the stabilizers of paths of length s are p'-groups.

Proof: By (12.5) and (12.6) we have $|S| \geq q^{11}$, and (12.7) and (12.8) imply $7 \leq s \leq 9$.

Let $\tilde{\gamma} = (\alpha,\beta,\delta_2,\ldots,\delta_7)$ be a path of length 7. By (6.7) we may assume that $K \leq G_{\tilde{\gamma}}$. Since $s \geq 7$ we get from (3.4) that $|S/S \cap G_{\tilde{\gamma}}| = q^9$ and so

$$|S \cap G_{\tilde{\gamma}}| \geq q^2.$$

If $|S \cap G_{\tilde{\gamma}}| \geq q^3$, then $s = 9$. Assume for a moment that $|S \cap G_{\tilde{\gamma}}| < q^3$. There are vertices $\delta_8 \in \Delta(\delta_7) \setminus \{\delta_6\}$ and $\delta_9 \in \Delta(\delta_8) \setminus \{\delta_7\}$ which are stabilized by K. Hence, the operation of K implies that $|G_{\tilde{\gamma}} \cap S/G_{\tilde{\gamma}} \cap Q_{\delta_k}| = q^i$ for $k \in \{8,9\}$ and some $i \in \mathbb{N}$. It follows that

$$|S \cap G_{\tilde{\gamma}}| = q^2 \text{ and } |S| = q^{11}.$$

Now (12.6) implies that V_β and Q_β/V_β are faithful irreducible L_β/Q_β-modules. It is easy to check that $V_{\alpha'}$ operates quadratically on these modules. Hence, (12.5), (5.11) and (5.18) yield $q = 2$.

Let $\hat{\gamma} = (\alpha,\beta,\ldots,\delta_8)$ be a path of length 8 so that $\tilde{\gamma} \subseteq \hat{\gamma}$ and let $Y = V_{\delta_3} \cap V_{\delta_5}$. As we have seen above $|G_{\tilde{\gamma}}| = 4$. On the other hand, by (12.6) Y is elementary abelian of order 8 and $Y \cap Q_\beta \leq G_{\tilde{\gamma}}$. It follows that $Y \cap Q_\beta = G_{\tilde{\gamma}}$. Hence, $G_{\tilde{\gamma}}$ is elementary abelian and cannot be transitive on $\Delta(\delta_7) \setminus \{\delta_6\}$; i.e. $s = 7$ and $r = 8$.

It remains to describe the chief factors of G_α and G_β. For G_β this was done in (12.5) and (12.6). We only need to note that for $s = 7$ (and thus $q = 2$) we have $G_\beta^{(2)} = V_\beta$.

We now describe the chief factors of G_α. For this purpose we define for fixed $\alpha-1 \in \Delta(\alpha) \setminus \{\beta\}$:

$H = V_\beta \cap V_{\alpha-1}$,

$V = V_\alpha^{(2)}$,

$W = \langle (V_\beta \cap Q_{\alpha-1})^{G_\alpha} \rangle$.

Clearly, we have $H \leq W \leq V$, and it is easy to check that $H = G_\alpha^{(3)}$. From (12.1) and (12.6)(c) we get $L_\alpha/Q_\alpha \simeq L_2(q)$ and the module structure of H. In addition, (12.1)(g) implies $[V_\beta \cap Q_{\alpha-1}, V_{\alpha-1} \cap Q_\beta] = 1$. Since G_α is doubly transitive on $\Delta(\alpha)$ it follows from the latter indentity that

(1) W is elementary abelian.

Since $|V_\beta/V_\beta \cap W| = q = |V_\beta/C_{V_\beta}(W)|$ we have as an immediate consequence that

(2) $W \leq G_\alpha^{(2)}$.

If $s = 7$, then $V_\beta = G_\beta^{(2)}$, and if $s = 9$, then $G_\beta^{(2)} = V_\beta(G_{\tilde{\gamma}} \cap Q_\alpha)$ where $G_{\tilde{\gamma}} \cap Q_\alpha$ is isomorphic to a Sylow 2-subgroup of $Sz(q)$. It follows that $\Omega_1(G_\beta^{(2)}) = V_\beta$; in particular, $[W,Q_\beta] \leq V_\beta \cap W$. Now the action of L_β on Q_β yields

(3) $[Q_\beta, Q_\beta] = V_\beta$.

In addition, $[V_{\alpha'} \cap Q_\beta, W] \leq W \cap V_\beta$, and since $V_{\alpha'} \cap Q_\beta \nleq Q_\alpha$ it follows that $(V_{\alpha-1} \cap Q_\beta)(V_\beta \cap Q_{\alpha-1})$ is normal in L_α. We get:

(4) W/H is a natural $SL_2(q)$-module.

Note that $|Q_\alpha/C_{Q_\alpha}(H)| = q^2$. It is easy to check that

(5) $Q_\alpha/C_{Q_\alpha}(H)$ is a natural $SL_2(q)$-module.

It remains to investigate the module structure of $C_{Q_\alpha}(H)/W = E$. Note that $|E| = q^4$ or $|E| = 2^3$. Furthermore, if E is an irreducible $SL_2(q)$-module, then $G_\alpha^{(2)} = W$.

Assume first that $V = V_{\alpha-1}V_\beta$. Then $[V_{\alpha-1}, V_{\alpha'} \cap Q_\beta] \le V_\beta$, and $(V_{\alpha'} \cap Q_\beta)V_\beta$ is normalized by $L = \langle V_{\alpha-1}, V_{\alpha'}\rangle$. But L is transitive on $\Delta(\beta)$, contradicting (12.6). Thus, we have

(6) $|V/W| > q^2$.

Assume now that $q = 2$. Let d be an element of order 3 in G_α which is inverted by some $v \in (V_{\alpha'} \cap Q_\beta) \setminus Q_\alpha$, and let $Z = C_V(d)$. By (6) we have $|V/W| = 2^3$, and (4) implies $[Z \cap V, v] \le Z \cap H \le V_\beta$. Hence, $\langle v\rangle V_\beta$ is normalized by $\langle Z, V_{\alpha'}\rangle$ and as above $Z \cap V \le Q_\beta$. Now the operation of d yields $W(Z \cap V) = G_\alpha^{(2)} \cap V$.

Set $Q = [Q_\alpha, d]$. Then (5) implies $|Q/Q \cap W| = 4^2$ and $Q_\alpha = ZQ$; in addition, $Z = Z \cap V$ if $s = 7$. Assume that $Q/Q \cap W$ is elementary abelian. Then $|Q/Q \cap Q_\beta| = 2$ and $[Q \cap Q_\beta, v] \le V_\beta \cap Q$ by (3). It follows that $[Q,v] = [V_{\alpha-1}(Q \cap Q_\beta), v] \le V$, and $Q/Q \cap V$ is centralized by v, a contradiction to (5). Hence, we have shown that $Q/Q \cap W \simeq C_4 \times C_4$. If $Z = Z \cap V$, then G is of type ${}^2F_4(2)'$, and if $Z \neq Z \cap V$, then G is of type ${}^2F_4(2)$.

Assume now that $q > 2$; i.e. $K_\alpha \neq 1$. Note that $K_\alpha \not\le G_{\alpha+3}^{(1)}$ since $[Z_\alpha, K_\alpha] \not\le Z_\alpha \cap Q_{\alpha+3}$, $K_\alpha \not\le G_{\alpha+2}^{(1)}$ since $[W, K_\alpha] \not\le Q_{\alpha+2}$ by (4), and $K_\alpha \not\le G_\beta^{(1)}$ since $[Q_\alpha, K_\alpha] \not\le C_{Q_\alpha}(H)$ by (5). It follows that $C_{Q_\alpha}(K_\alpha)$ stabilzes a path of length 8 and has order q; i.e. $C_{Q_\alpha}(K_\alpha) \le H$.

To prove assertion (c), it remains to show that E is irreducible. Let $\bar{E}_o$ be an irreducible L_α-submodule in E, and let E_o be its preimage. Since $C_E(K_\alpha) = 1$ we get $|\bar{E}_o| = q^2 = |E/\bar{E}_o|$, and (6) yields $VE_o = C_{Q_\alpha}(H)$.

Note that $[E_o \cap Q_\beta, V_{\alpha'} \cap Q_\beta] \leq V_\beta \cap W$ by (3) and that $[V_{\alpha-1}, V_{\alpha'} \cap Q_\beta] \leq V_\beta E_o$. We conclude that $E/\bar{E}_o$ and $\bar{E}_o$ are natural $SL_2(q)$-modules. It follows together with (5) and (5.17) that Q_α/W is elementary abelian. But this contradicts the structure of $Q_\alpha/Q_\alpha \cap Q_\beta$. □

Remark: (1) In handling the case $q = 2$ in (12.9), i.e. G is of type ${}^2F_4(2)'$ or ${}^2F_4(2)$, we determined more than just the chief factors of G_α and G_β, namely, we were able to identify the chief factors in terms of certain natural "geometric" subgroups. These data should be able to serve as a starting point for determining generators and relations for the given groups and concluding that G is, in fact, locally isomorphic to ${}^2F_4(2)'$ or ${}^2F_4(2)$. This analysis has been carried through by P. Fan in [5].

(2) In case (b) of (12.9), i.e. $r = s-1 = 8$, and $q = 2$, G contains a "subamalgam" of type ${}^2F_4(2)'$. We leave the details as an exercise to the reader (see chapter 16).

13. The case $[V_{\alpha'}, V_\beta \cap Q_{\alpha'}] = 1$.

In this chapter we conclude the analysis of the structure of the groups G_α and G_β. The only case we have as yet not completely treated is that considered in chapter 9. Hence, we are allowed to set:

(13.0) Hypothesis. In this chapter Hypothesis (9.0) holds.

We will show that only three exceptional cases arise under this hypothesis, namely when G is parabolic isomorphic to M_{12}, $\mathrm{Aut}(M_{12})$ or is of type F_3.

(13.1) Assume that q_α is even. Then

(a) $q_\alpha = q_\beta = 2$,

(b) $G_\beta/Q_\beta \simeq SL_2(2)$ and $L_\delta = G_\delta$, $\delta \in \Gamma$,

(c) $b = 3$,

(d) $|Z_\beta|^2 = |Z_\alpha| = 4$, $Z(L_\alpha) = 1$,

(e) $V_\beta = Z_\alpha Z_{\alpha+2}$, $|V_\beta| = 8$,

(f) $G_\beta/Q_\alpha \cap Q_{\alpha+2} \simeq \Sigma_4$.

Proof: Apart from (f) this is (9.1), (9.3), (9.4), (9.8), (9.10), and (9.11).

To prove (f) we observe that $C_{Q_\beta}(V_\beta) = Q_\alpha \cap Q_{\alpha+2}$ is normal in G_β and has index at most 4 in Q_β. Since by (6.5) $Q_\alpha \cap Q_\beta$ is not normal in G_β, we get (f). □

The result (13.1) will frequently be used without reference.

(13.2) <u>Assume that</u> q_α <u>is even.</u> <u>Let</u> $\tilde{\gamma} = (\delta_0,\ldots,\delta_6)$ <u>be a path of length</u> 6 <u>with</u> $\delta_0 \in \beta^G$. <u>Then</u> $[Z_{\delta_0}, Z_{\delta_6}] = Z_\mu$ <u>where</u> $\mu \in \Delta(\delta_3) \setminus \{\delta_2,\delta_4\}$.

<u>Proof</u>: By (6.8) $s \geq 4$, so we may assume without loss that $(\delta_1,\delta_2,\delta_3,\delta_4) = \gamma$. It follows that $[Z_{\delta_0}, Z_{\delta_6}] = [V_\beta, V_{\alpha'}]$, and (13.1)(e) implies that $[V_\beta, V_{\alpha'}] \leq Z_{\alpha+2}$. If $[V_\beta, V_{\alpha'}] = Z_{\alpha'}$, then V_β centralizes $V_{\alpha'}/Z_{\alpha'}$, contradicting (3.2)(c). With the same argument $[V_\beta, V_{\alpha'}] \neq Z_\beta$ and so $[V_\beta, V_{\alpha'}] = Z_\mu$, $\mu \in \Delta(\alpha+2) \setminus \{\beta,\alpha'\}$.

(13.2) <u>Assume that</u> q_α <u>is even.</u> <u>Then</u> G <u>is parabolic isomorphic to</u> M_{12} <u>or</u> $\mathrm{Aut}(M_{12})$.

<u>Proof</u>: We choose the following notation for $\delta \in \beta^G$ and $\lambda \in \Delta(\delta)$:

$$Q^*_\lambda = V_\lambda^{(2)},$$

$$Q^*_\delta = \langle Q^*_\lambda \cap Q_\delta \mid \lambda \in \Delta(\delta)\rangle,$$

$$G^*_\lambda = \langle Q^{*\,G_\lambda}_\delta\rangle,$$

$$G^*_\delta = \langle Q^{*\,G_\delta}_\lambda\rangle,$$

$$S^* = Q^*_\alpha Q^*_\beta,$$

$$G^* = \langle G^*_\alpha, G^*_\beta\rangle.$$

Note first that $|Q^*_\alpha/Z_\alpha| \leq 8$ by (13.1)(e), and since Q^*_α is non-abelian we have $|Q^*_\alpha/Z_\alpha| = 8$ and after conjugation

(1) $|Q^*_\lambda| = 2^5$.

In particular, $|Q^*_\alpha \cap Q_\beta/V_\beta| = 2$ and from $[Q^*_\alpha, Q_\beta] \leq Q^*_\alpha \cap Q_\beta$ we conclude that

(2) $Q^*_\beta = (Q^*_\alpha \cap Q_\beta)(Q^*_{\alpha+2} \cap Q_\beta)$ and $|Q^*_\delta| = 2^5$.

It is now easy to check that

(3) G^* is an amalgam of G^*_α and G^*_β over S^*.

Note that M_{12} fulfils Hypothesis (13.0) for two subgroups P_1 and P_2 containing a common Sylow 2-subgroup of M_{12}. In addition $|P_i| = 3 \cdot 2^6 = |G^*_\alpha|$ and so M_{12} is an example for such a group G^*.

Our plan of action is to describe the subgroups G^*_α and G^*_β by generators and relations. This will show that G^* is parabolic isomorphic to M_{12}.

For $\delta \in \beta^G$ we set $Z_\delta = \langle t_\delta \rangle$, so we have $t_\delta^2 = 1$. Now choose $\langle d \rangle \in Syl_3(G_\alpha)$ and set $\beta^d = \alpha-1$. Then $G^*_\alpha = \langle t_\beta, t_{\alpha-1}, t_{\alpha'}, t_{\alpha'}^d, t_{\alpha'}^{d^{-1}} \rangle$ and (13.2) allows us to compute the commutators:

(4) $[t_{\alpha'}^{d^i}, t_{\alpha'}^{d^j}] = t_\beta^{d^k}$ for $\{i,j,k\} = \{1,2,3\}$.

Hence, (4) together with the other obvious relations determines the isomorphism type of $Q^*_\alpha\langle d \rangle$. In particular, we can describe $C_{Q^*_\alpha}(d)$. Obviously, $C_{Q^*_\alpha}(d) = \langle z \rangle$ for some involution z, and it is easy to check that

(5) $z = t_{\alpha'} t_{\alpha'}^d t_{\alpha'}^{d^{-1}} t_{\alpha-1}$,

and

(6) $[z,t_{\alpha'}] = t_\beta$.

Now (6) shows that z centralizes V_β/Z_β and $z \in Q^*_\beta$ but $z \notin Q^*_{\alpha+2}$. Hence, there exists $<\tilde{d}> \in Syl_3(G^*_{\alpha+2})$ so that $\beta^{\tilde{d}} = \alpha'$ and

(7) $\tilde{d}^z = \tilde{d}^{-1}$.

We set $C_{Q^*_{\alpha+2}}(\tilde{d}) = <\tilde{z}>$. It follows that

(8) $[z,\tilde{z}] = 1$.

As in (5) we deduce that

(9) $\tilde{z} = t_{\alpha-1}t_{\alpha-1}^{\tilde{d}}t_{\alpha-1}^{\tilde{d}^{-1}}t_{\alpha'}$ and $[\tilde{z},t_{\alpha-1}] = t_\beta$.

Note that $G^*_\alpha/<z>Z_\alpha \simeq \Sigma_4$ and $C_{Q^*_\alpha}(z) = <z>Z_\alpha$. Hence, by (8) $\tilde{z}$ normalizes $<d>Z_\alpha$. Since any element in dZ_α has the same properties as d in the above relations, we may rearrange notation so that

(10) $d^{\tilde{z}} = d^{-1}$.

This shows that $<Q^*_\alpha,d,\tilde{z}> = G^*_\alpha$ is uniquely determined.

It remains to determine G^*_β, where we already have relations for $Q^*_\beta = <t_\beta,t_{\alpha-1},t_{\alpha'},z,\tilde{z}>$. Note first that $t_{\alpha'}^d \notin Q_\beta$ and G^*_β/Z_β contains no element of order 6. Hence, there exists $t_\sigma \in \{t_{\alpha-1}^{\tilde{d}},t_{\alpha-1}^{\tilde{d}}t_\beta\} \subseteq V_\beta^{\tilde{d}}$ $(= V_{\alpha'})$ so that

(11) $\langle t_{\alpha'}^{d}, t_{\sigma}\rangle \simeq \Sigma_3$, i.e. $(t_{\alpha'}^{d} t_{\sigma})^3 = 1$.

Since $d^{\tilde{z}} = d^{-1}$ we get $[t_{\alpha'}^{d}, \tilde{z}] = t_{\alpha'}^{d} t_{\alpha'}^{d^{-1}}$ and then from (5)

(12) $[t_{\alpha'}^{d}, \tilde{z}] = t_{\alpha'} z t_{\alpha-1}$.

Similarly, we get from (9) that

(13) $[t_{\sigma}, z] = t_{\alpha-1} \tilde{z} t_{\alpha'}$.

Note here that (13) does not depend on the choice of t_{σ} in $\{t_{\alpha-1}^{\tilde{d}}, t_{\alpha-1}^{\tilde{d}} t_{\beta}\}$ since t_{β} centralizes both t_{σ} and z.

From these last two relations we conclude that $G_{\beta}^{*} = \langle t_{\alpha'}^{d}, t_{\sigma}, Q_{\beta}^{*}\rangle$ is uniquely determined, and G^{*} is parabolic isomorphic to M_{12}.

If $G_{\delta} = G_{\delta}^{*}$ for $\delta \in \Gamma$, then G is parabolic isomorphic to M_{12} and we are done. So we assume for the remainder of the proof that $G_{\delta} \neq G_{\delta}^{*}$. We will show that G is parabolic isomorphic to $\mathrm{Aut}(M_{12})$. This will be done by the same method as above. We will show that the isomorphic type of G_{α} and G_{β} is uniquely determined, and since $\mathrm{Aut}(M_{12})$ satisfies the hypothesis with $G_{\delta} \neq G_{\delta}^{*}$, then G is parabolic isomorphic to $\mathrm{Aut}(M_{12})$.

Of course, we adopt the notation of the first part of the proof. In addition, we set $D_{\alpha} = \bigcap_{\delta \in \Delta(\alpha)} Q_{\delta}$ and $C = C_{Q_{\alpha}}(d)$.

(14) $D_{\alpha} = \langle z\rangle Z_{\alpha}$.

We have already shown that $\langle z\rangle Z_{\alpha} \leq D_{\alpha}$. Assume that $D_{\alpha} \neq \langle z\rangle Z_{\alpha}$.

Since $Q_\alpha = CQ^*_\alpha$ we get that $D_\alpha \cap C \neq \langle z\rangle$; in particular $C_0 = C \cap D_\alpha \cap Q_{\alpha+2} \neq 1$. It follows from (13.1)(e) that $[C_0,V_\beta] = 1$ and thus $[C_0,Q^*_\alpha] = 1$. Hence, C_0 is normal in $Q_\alpha\langle d\rangle$ and $Z(L_\alpha) \neq 1$, a contradiction.

(15) $G_\alpha/D_\alpha \simeq C_2 \times \Sigma_4$.

Clearly, $|Q_\alpha/D_\alpha| \leq 8$ and by (14) $D_\alpha \leq Q^*_\alpha$. Since $Q_\alpha \neq Q^*_\alpha$ the assertion follows.

(16) CZ_α is elementary abelian.

Note that by (14) and (15) CZ_α is normal in G_α and $C \cap Z_\alpha = 1$. Hence, $\phi(CZ_\alpha) = \phi(C)$ and $\phi(C) \cap \Omega_1(Z(S)) = 1$ imply the assertion.

We now choose $c \in C \setminus \langle z\rangle$. Then c is an involution and $c \notin S^*$. In addition, by (15) we have $\langle c,G^*_\alpha\rangle = G_\alpha$ and $\langle c,G^*_\beta\rangle = G_\beta$. We have to find the relations of c with the above generators of G^*_α and G^*_β. Clearly,

(17) $[c,t_{\alpha-1}] = [c,t_\beta] = [c,d] = [c,z] = 1$.

By (14) $c \notin Q_\beta$ and thus $[c,t_{\alpha'}] = t_{\alpha-1}$ or $t_{\alpha-1}t_\beta$. Since $[cz,t_{\alpha'}] = [c,t_{\alpha'}]t_\beta$ we may choose c without changing the relations in (17) so that

(18) $[c,t_{\alpha'}] = t_{\alpha-1}$.

Next we observe that $[c,\tilde{z}] = 1$ or z since $\tilde{z}$ normalizes $\langle d\rangle$.

If $[c,\tilde{z}] = 1$, then c centralizes $\langle z,\tilde{z}\rangle$ and thus $Q_\beta/Q_\alpha \cap Q_{\alpha+2}$ which contradicts $c \notin Q_\beta$ and $G_\beta/Q_\alpha \cap Q_{\alpha+2} \simeq \Sigma_4$. So we have

(19) $[c,\tilde{z}] = z$.

Note that $\langle c,t_\sigma\rangle = H$ is a dihedral group whose order is divisible by 3. Set $D_\beta = Q_\alpha \cap Q_{\alpha+2}$ and $H_o = H \cap Q_\beta$. Since H_o centralizes an element of order 3 in H, we get again from $G_\beta/D_\beta \simeq \Sigma_4$ that $H_o \leq D_\beta$.

If $H_o = 1$, then $H \simeq \Sigma_3$ and $c \in t_\sigma^H \subseteq G_\beta^*$, a contradiction. So $H_o \neq 1$, and we assume first that $H_o \leq CZ_\alpha$. Since by (15) $G_\alpha/CZ_\alpha \simeq \Sigma_4$ and $Z(L_\alpha) = 1$, every element in $CZ_\alpha \setminus Z_\alpha$ is centralized by an element of order 3. Hence, (3.2)(c) implies $H_o \leq Z_\alpha$ and then the operation of H on V_β/Z_β that $H_o = Z_\beta$. But now $c \in t_\sigma^H \cup (t_\sigma t_\beta)^H \subseteq G_\beta^*$, again contradicting $c \notin G_\beta^*$.

We have shown that $H_o \nleq CZ_\alpha$. If $[H_o,c] = 1$, then by (15) and (16) $c \in Z(Q_\alpha)$ and $Z(L_\alpha) \neq 1$, a contradiction. Hence, $[H_o,c] \neq 1$ and since Q_α has exponent 4 we conclude that $H_o \simeq C_4$ and $H \simeq D_{24}$ and

(20) $(ct_\sigma)^{12} = 1$.

This last relation completes the set of relations between the generators

$$t_{\alpha-1}, t_\beta, t_{\alpha'}, d, \tilde{z} \text{ and } c \text{ for } G_\alpha$$

and

$$t_\beta, t_{\alpha-1}, t_{\alpha'}, z, \tilde{z}, t_{\alpha'}^d, t_\sigma \text{ and } c \text{ for } G_\beta,$$

and G_α and G_β are uniquely determined. □

We finish the case q_α even by determining the parameters r,s for the group G.

(13.4) <u>Assume that</u> q_α <u>is even. Then either</u>

(a) G <u>is parabolic isomorphic to</u> M_{12} <u>and</u> $s = 5$ <u>and</u> $r = 8$, <u>or</u>

(b) G <u>is parabolic isomorphic to</u> $\mathrm{Aut}(M_{12})$ <u>and</u> $s = 7$ <u>and</u> $r = 8$.

<u>Proof</u>: Note that by (3.4)(c) any two paths of lengths $d \leq s$ with conjugate initial vertices are conjugate, and note further that by (6.8) $s \geq 4$.

Let $\gamma_1 = (\delta_o,\ldots,\delta_d)$ be a path of even length d with $\delta_o \in \beta^G$. If $d = 8$, then $\delta_4 \in \beta^G$ and $Z_{\delta_4} \leq G_{\gamma_1}$, and Z_{δ_4} operates transitively on $\Delta(\delta_o) \setminus \{\delta_1\}$ and $\Delta(\delta_d) \setminus \{\delta_{d-1}\}$ since $b = 3$. It follows that all paths γ_1 of even length $d \leq 8$ with $\delta_o \in \beta^G$ are regular, in particular $r \geq 8$.

Now let $\gamma_2 = (\alpha-2,\alpha-1,\alpha,\beta,\alpha+2)$ be a path of length 4 and pick $z \in Q^*_\alpha$ as in (13.3). Then $z \notin Q_{\alpha+2}$ (see steps (5) - (7) in the proof of (13.3)) and with the same argument $z \notin Q_{\alpha-2}$. Hence, γ_2 is regular and $s \geq 5$.

Now let $\gamma_3 = (\alpha-2,\alpha-1,\alpha,\beta,\alpha+2,\alpha',\alpha'+1)$ be a path of length 6. Set $D_\beta = Q_\alpha \cap Q_{\alpha+2}$ and $D = D_\beta \cap Q_{\alpha-1} \cap Q_{\alpha'}$. Since $b = 3$ we have

$$D_\beta = V_\beta D \quad\text{and}\quad \phi(D_\beta) = \phi(D).$$

Assume first that G is parabolic isomorphic to M_{12}. Then (13.3) yields $D_\beta = V_\beta$ and $D_\beta \cap Q_{\alpha-1} = Z_\alpha \leq Q_{\alpha-2}$. Hence, the path $(\alpha-2,\alpha-1,\alpha,\ldots,\alpha')$ is not regular and $s = 5$. Since $|G_\alpha| = 3\cdot 2^6$ it is easy to check that (a) holds.

Assume now that G is parabolic isomorphic to $Aut(M_{12})$. Using (13.3) and the notation there, we get $D_\beta = H_o V_\beta$ and $D_\beta \simeq C_4 \times C_2 \times C_2$; in particular $\phi(D_\beta) = Z_\beta$. Thus, if $D \leq Q_{\alpha-2}$, then $D \leq Q_{\alpha-2} \cap Q_\alpha$ and $\phi(D) = \phi(D_\beta) = Z_\beta \leq \phi(Q_{\alpha-2} \cap Q_\alpha) = Z_{\alpha-1}$, a contradiction. With the same argument we have $D \not\leq Q_{\alpha'+1}$. It follows that $s \geq 7$. Since $|G_\alpha| = 3 \cdot 2^7$ we conclude that (b) holds. □

Since we have already treated the case where q_α is even, we assume now:

Hypothesis. In the following q_α is odd.

(13.5)(a) $q_\alpha = q_\beta = 3$.

(b) $b = 5$.

(c) $L_\delta/Q_\delta \simeq SL_2(3)$, $\delta \in \Gamma$.

(d) Z_α is a natural L_α/Q_α-module; in particular $Z(L_\alpha) = 1$.

(e) $|Z_\beta|^2 = |Z_\alpha| = 9$.

(f) $V_\beta = Z_\alpha Z_{\alpha+2}$ and V_β/Z_β is a natural L_β/Q_β-module.

(g) $s \geq 4$.

Proof: This is (9.3), (9.4), (9.8), (9.10), (9.11), and (6.8). □

We will use (13.5) mostly without reference.

(13.6) Let $\tilde{\gamma} = (\delta_o, \ldots, \delta_4)$ be a path of length 4 so that $\delta_o \in \beta^G$. Then $\tilde{\gamma}$ is regular and $[V_{\delta_o}, V_{\delta_4}] = Z_{\delta_2}$.

Proof: Since $s \geq 4$ we may assume that $\delta_0 = \beta$ and $\delta_4 = \alpha'$, and since $V_\beta \not\leq Q_{\alpha'}$, and by (9.2) $V_{\alpha'} \not\leq Q_\beta$ the path $\tilde{\gamma}$ is regular. The second assertion follows as (*) in the proof of (9.11). □

(13.7) There exists a $(G_\alpha \cap G_\beta)$-conjugate of K stabilizing γ.

Proof: By (13.6) $(\alpha',\ldots,\beta)$ is regular and thus by (6.7) there exists $K^g \leq G_\gamma$ $(g \in G)$. Since $K, K^g \leq G_\alpha \cap G_\beta$ and $|G_\alpha \cap G_\beta| = 4 \cdot 3^*$, the assertion follows from Sylow's theorem. □

Notation. In view of (13.7) we may assume that $K \leq G_\gamma$. In addition, we set

$$W_\delta = V_\delta^{(4)}, \qquad \delta \in \beta,$$

$$\langle t_\delta \rangle = K \cap L_\delta, \qquad \delta \in \gamma.$$

Note that t_δ is an involution contained in $G_\delta^{(1)}$.

(13.8)(a) $G_\beta^{(2)} = Q_\alpha \cap Q_{\alpha+2}$, and $Q_\beta/G_\beta^{(2)}$ is a natural L_β/Q_β-module.

(b) $[V_\alpha^{(2)}, Q_\alpha] = [V_\alpha^{(2)}, t_\alpha] = Z_\alpha$.

(c) $Q_\alpha/G_\alpha^{(3)}$ is elementary abelian.

Proof: Since $V_\beta = Z_\alpha Z_{\alpha+2}$ we have $C_{Q_\beta}(V_\beta) = Q_\alpha \cap Q_{\alpha+2}$; in particular $Q_\alpha \cap Q_{\alpha+2}$ is normal in G_β and $G_\beta^{(2)} = Q_\alpha \cap Q_{\alpha+2}$. Note that $|Q_\beta/G_\beta^{(2)}| \leq 9$ and by (6.5) $G_\beta^{(2)} \neq Q_\alpha \cap Q_\beta$. Hence, $|Q_\beta/G_\beta^{(2)}| = 9$ and (a) follows.

By (6.9) $t_\alpha \neq t_\beta$ and thus t_α acts fixed-point-freely on Z_α but not on V_β; i.e. $[V_\beta, t_\alpha] = Z_\alpha$. Since $C_{L_\alpha}(t_\alpha)$ is transitive on $\Delta(\alpha)$, we get $[V_\alpha^{(2)}, t_\alpha] = Z_\alpha$. As $1 \neq [V_\beta, Q_\alpha] \leq Z_\alpha$ we have $1 \neq [V_\alpha^{(2)}, Q_\alpha] \leq Z_\alpha$, and as $[V_\alpha^{(2)}, Q_\alpha]$ is normal in L_α, we get $[V_\alpha^{(2)}, Q_\alpha] = Z_\alpha$.

Note that by (b) $[V_\alpha^{(2)}, Q_\alpha, Q_\alpha] = 1$. Hence (5.8) implies that $Q_\alpha/C_{Q_\alpha}(V_\alpha^{(2)})$ is elementary abelian. But clearly, $C_{Q_\alpha}(V_\alpha^{(2)}) = G_\alpha^{(3)}$ and (c) follows. □

(13.9) $Z(L_\alpha/Z_\alpha) = 1$.

Proof: Let L_α be a counter example and $Z_\alpha \leq C \leq L_\alpha$ so that $|C/Z_\alpha| = 3$ and $C/Z_\alpha \leq Z(L_\alpha/Z_\alpha)$. Set $Z = C_C(t_\alpha)$, then $|Z| = 3$ and $Z \cap Z_\alpha = 1$.

Note that by (6.9) $t_\beta = t_{\alpha+4}$ and $t_\alpha = t_\delta$ for $\delta \in \gamma$, if and only if $\delta = \alpha+3$. It follows that $Z \leq Q_\beta \cap Q_{\alpha+2}$.

Assume first that $Z \leq Q_{\alpha+3}$. Then $Z \leq Q_{\alpha+3} \cap Q_{\alpha+4} \cap Q_{\alpha'}$, since $t_{\alpha+4} \neq t_\alpha \neq t_{\alpha'}$, and

$$[V_{\alpha'}, Z] \leq Z_{\alpha'}.$$

On the other hand, $[Z, V_{\alpha+3}] = 1$ and since $s \geq 4$ and C is normal in L_α we get $C \leq Z(W_\beta)$. But W_β is non-abelian and $Z_{\alpha+4} \cap Z(W_\beta) = Z_{\alpha+3}$. It follows that $[V_{\alpha'}, Z] \leq Z_{\alpha'} \cap Z(W_\beta) = 1$. By transitivity on 3-paths $[V_{\alpha-3}, C] = 1$ for $d(\alpha-3, \alpha) = 3 = d(\beta, \alpha-3)-1$, and $\langle V_{\alpha-3}, V_{\alpha'} \rangle$ is transitive on $\Delta(\beta)$. Now (3.2)(c) applied to $C_G(Z)$ yields a contradiction.

Assume now that $Z \not\leq Q_{\alpha+3}$. Set $T = Q_{\alpha+2}Q_{\alpha+3}$ and $\bar{L}_{\alpha+3} = L_{\alpha+3}/G_{\alpha+3}^{(2)}$.

Since $t_\alpha = t_{\alpha+3}$ we get from (13.8)(a) that $|C_{\bar{T}}(t_\alpha)| = 3$ and $\bar{Z} = C_{\bar{T}}(t_\alpha)$. On the other hand, since $s \geq 4$ there exists $\alpha-1 \in \Delta(\alpha)$ so that $K \leq G_{\alpha-1}$ and $V_{\alpha-1} \not\leq Q_{\alpha+3}$. It follows that $\bar{Z} = \overline{C_{V_{\alpha-1}}(t_\alpha)}$, and we get:

$$[W_{\alpha+3},Z] \leq [W_{\alpha+3},V_{\alpha-1}][W_{\alpha+3},G_{\alpha+3}^{(2)}] \leq [W_{\alpha+3},V_{\alpha-1}]V_{\alpha+3} \leq W_\beta.$$

Hence, ZW_β is normal in $\langle S,V_{\alpha'}\rangle = G_\beta$. Since $t_\beta \neq t_{\alpha+3}$ Z is inverted by t_β and so $Z \leq [W_\beta,t_\beta]$. We now determine $[W_\beta,t_\beta]$.

We have $|Q_\alpha/C_{Q_\alpha}(C)| = 9$ and so $W_{\alpha+3} \cap Q_\beta \leq V_{\alpha+3}C_S(t_\alpha)C_{Q_\alpha}(C)$ and

$$[W_{\alpha+3} \cap Q_\beta,Z] = [V_{\alpha+3},Z] = Z_\alpha \cap Z_{\alpha+2} = Z_\beta.$$

Set $W = W_{\alpha+3}/V_{\alpha+3}$, then $|W/C_W(Z)| = 3$ and thus $|[W,t_{\alpha+3}]| = 9$. Hence, also $|[W_\beta,t_\beta]V_\beta/V_\beta| = 9$, and since t_β inverts $V_{\alpha-1}/Z_\alpha$ and $V_{\alpha+3}/Z_{\alpha+2}$ we have

$$[W_\beta,t_\beta]V_\beta = V_{\alpha-1}V_{\alpha+3}V_\beta = ZV_{\alpha+3}V_\beta.$$

But now $ZV_\beta = C_{[W_\beta,t_\beta]}(t_\alpha)V_\beta = V_{\alpha-1}V_\beta$, and S normalizes $V_{\alpha-1}V_\beta$. Hence, $V_{\alpha-1}V_\beta = V_\alpha^{(2)}$ and $[V_\alpha^{(2)},O^3(L_\alpha)] \leq Z_\alpha$, contradicting (3.2)(c) applied to $N_G(V_\beta)$. □

(13.10)(a) $Z_\beta = \phi(W_\beta) = W_\beta'$ <u>and</u> $[W_\beta,Q_\beta] = V_\beta$.

(b) $|V_\alpha^{(2)}| = 3^5$.

(c) $Z(W_\beta)$ <u>is elementary abelian of order</u> 3^4.

(d) $W_\beta/Z(W_\beta)$ <u>is a natural</u> L_β/Q_β<u>-module</u>.

<u>Proof</u>: Note that $[V_\alpha^{(2)},Q_\beta] \nleq Z_\alpha$. Hence, (a) is an easy consequence of (13.6) and (d).

Set $V = V_\alpha^{(2)}/Z_\alpha$. By (13.8)(b) V is an $L_2(3)$-module, and V is generated by 4 subgroups of order 3, namely V_δ/Z_δ for $\delta \in \Delta(\alpha)$. It follows that $|V| \leq 3^4$. In addition, since V_β is not normal in L_α, we have $[V,O^3(L_\alpha)] \neq 1$ and by (5.9) S cannot operate quadratically on V. This shows that $|V| \geq 3^3$. Assume that $|V| = 3^4$. Then V is an A_4-permutation module $(A_4 \simeq L_2(3))$ and $C_V(L_\alpha) \neq 1$, contradicting (13.9). Hence, (b) follows.

Let $R = C_{V_\alpha^{(2)}}(K)$. Then $|R| = 3$ and $R \leq G_\gamma$ by (6.9). Thus, RV_β is normal in $\langle S,V_{\alpha'}\rangle = G_\beta$; in particular $RV_\beta \leq \Omega_1(Z(W_\beta))$. Set $W = W_\beta/RV_\beta$. Since $|V_\alpha^{(2)}/RV_\beta| = 3$ we have $|W| \leq 3^4$ and from (3.2)(c) $|W| \geq 3^2$.

Note that by (a) $[W,t_\beta]$ is normal in G_β/RV_β. On the other hand, by (6.9) $t_\beta = t_{\alpha+4}$ and so $Z_{\alpha+4} \leq [W_\beta,t_\beta]$. It follows that $W = [W,t_\beta]$. But $[W_\beta,V_{\alpha'},V_{\alpha'}] \leq [W_{\alpha+3},V_{\alpha'}] \leq Z_{\alpha+3} \leq V_\beta$ by (13.6), and $V_{\alpha'}$ operates quadratically on W. We conclude that $|W| = 3^2$, and (c) and (d) follow. □

(13.11)(a) $|S| = 3^{10}$.

(b) $|C_S(t_\alpha)| = 3^4$.

(c) $G_\alpha^{(3)} = V_\alpha^{(2)}$, <u>in particular</u> $V_\alpha^{(2)}$ <u>is elementary abelian</u>.

(d) $V_\alpha^{(2)}/Z_\alpha$ <u>is a non-faithful irreducible</u> L_α/Q_α<u>-module of order</u> 3^3.

(e) <u>There exists a normal series</u> $V_\alpha^{(2)} \leq N \leq Q_\alpha \leq L_\alpha$ <u>so that</u> $N/V_\alpha^{(2)}$ <u>and</u> Q_α/N <u>are natural</u> L_α/Q_α<u>-modules</u>.

<u>Proof</u>: By (13.10) $W_\beta = V_\alpha^{(2)} V_{\alpha+2}^{(2)}$ and $Z(W_\beta) = V_\alpha^{(2)} \cap V_{\alpha+2}^{(2)} \leq W_{\alpha+3}$. Hence, again by (13.10), $W_\beta \cap W_{\alpha+3} = V_{\alpha+3}Z(W_\beta)$ and $|W_{\alpha+3}/W_{\alpha+3} \cap W_\beta| = 3$; in particular $W_{\alpha+3} \cap Q_\beta = W_{\alpha+3} \cap W_\beta$.

Set $X = G_\alpha^{(3)}$. It follows that $[X, W_{\alpha+3}] \leq W_{\alpha+3} \cap Q_\beta \leq W_\beta$, and XW_β is normal in G_β and $[X, t_\beta] \leq W_\beta \cap X = V_\alpha^{(2)}$. Since $t_\alpha \in \langle t_\beta^{G_\alpha} \rangle$ and X and $V_\alpha^{(2)}$ are normal in G_α, we conclude that $[X, t_\alpha] \leq V_\alpha^{(2)}$ and thus $[X, t_\alpha] = Z_\alpha$ by (13.8)(b); in particular $X = C_X(t_\alpha)Z_\alpha$.

Set $Y = Q_\alpha/X$. By (13.8)(c) Y is elementary abelian and by (6.9) $C_{Q_\alpha}(t_\alpha) \leq G_{\alpha+3}$. We conclude from $s \geq 3$ that $C_{Q_\alpha}(t_\alpha) \leq X$ and $Q_\alpha = [Q_\alpha, t_\alpha]X$.

Note that $\phi(X) = \phi(C_X(t_\alpha))$ and $C_X(t_\alpha) \cap Z_\alpha = 1$. It follows that $\phi(X) = 1$. Hence $[Q_\alpha, X] = [Q_\alpha, t_\alpha, X]$, and the Three-subgroup Lemma yields $[Q_\alpha, X] = Z_\alpha$.

If $X \neq V_\alpha^{(2)}$, then there exists $x \in X \setminus V_\alpha^{(2)}$ so that $\langle x \rangle V_\alpha^{(2)}$ is normal in L_α, and L_α induces $L_2(3)$ on $\langle x \rangle V_\alpha^{(2)}/Z_\alpha$. It follows that $Z(L_\alpha/Z_\alpha) \neq 1$, contradicting (13.9). Thus, (c) holds.

As we have seen above $C_{Q_\alpha}(t_\alpha) \leq X = V_\alpha^{(2)}$. Now (13.10)(b) yields $|C_S(t_\alpha)| = 3^4$, and (b) holds. Now again (6.9) implies $|C_S(t_\alpha)| = |C_{Q_\alpha}(t_\beta)| = 3^4$.

As we have seen t_α operates fixed-point-freely on Y. Hence, $|Y| = |C_Y(t_\beta)|^2$. Since $|C_{V_\alpha^{(2)}}(t_\beta)| = 3^2$ we get that $|C_Y(t_\beta)| = 3^2$ and $|Y| = 3^4$. This shows together with (13.10)(b) that $|S| = 3^{10}$, and (a) holds.

By [2,VII 3.1] there is only one faithful irreducible $SL_2(3)$-module, respectively $L_2(3)$-module over $GF(3)$. Hence, (d) and (e) follow from $[Y,t_\alpha] = Y$ and (13.10)(a) and (13.9). □

(13.12) $s = 7$ <u>and</u> G <u>is of type</u> F_3.

<u>Proof</u>: The normal series

$$Z_\alpha \leq V_\alpha^{(2)} \leq N \leq Q_\alpha \leq L_\alpha \quad \text{(as in (13.11)(e))}$$

and

$$Z_\beta \leq V_\beta \leq Z(W_\beta) \leq W_\beta \leq G_\beta^{(2)} \leq Q_\beta \leq L_\beta$$

together with (13.5), (13.8), (13.10), and (13.11) show that G is of type F_3.

It remains to prove that $s = 7$. Note that $s \geq 4$ by (6.8). Let $\tilde{\gamma} = (\delta_0,\ldots,\delta_s)$ be a path of length s so that $G_{\tilde{\gamma}}$ is not transitive on $\Delta(\delta_0) \setminus \{\delta_1\}$. Then any 3-element in $G_{\tilde{\gamma}}$ is in $G_{\delta_0}^{(1)}$.

Assume first that $\delta_0 \in \beta^G$. Then $V_{\delta_4} \nleq G_{\tilde{\gamma}}$ since $V_{\delta_4} \nleq Q_{\delta_0}$ and so $s \geq 9$ in this case. Consider now the subpath $\hat{\gamma} = (\delta_1,\ldots,\delta_9)$ of length 8. By (3.4)(c) and (6.7) we may assume that $\gamma \subseteq \tilde{\gamma}$ and $K \leq G_{\tilde{\gamma}} \leq G_{\hat{\gamma}}$. Let $R = C_S(K)$. Then (13.11)(d) and (e) yield $|R| = 3$, and $R \leq G_{\tilde{\gamma}}$. Since $\hat{\gamma}$ is regular, we get $|O_3(G_{\hat{\gamma}})| = 3^3$. It follows that $O_3(G_{\hat{\gamma}}) = RZ_{\delta_5} \leq G_{\tilde{\gamma}}$, and $\hat{\gamma}$ is not regular, a contradiction.

Hence $\delta_0 \in \alpha^G$. By (3.4)(c) we may assume that $\delta_0 = \alpha$ and $\gamma \cap \tilde{\gamma} = (\alpha,\ldots,\delta_k)$ where $k = \min\{5,s\}$, and from (6.7) that $K \leq G_{\tilde{\gamma}}$. Since $t_\alpha = t_{\alpha+3}$ from (6.9) we deduce $C_S(t_\alpha) = C_{V_{\alpha-1}}(t_\alpha)C_{Q_{\alpha+3}}(t_\alpha)$ for $\alpha-1 \in \Delta(\alpha) - \{\beta\}$ with $K \leq G_{\alpha-1}$, and, again by (6.9), that

$C_{Q_{\alpha+3}}(t_\alpha) \leq Q_{\delta_i}$ for $1 \leq i \leq 5$. However, $C_{V_{\alpha-1}}(t_\alpha) \leq Q_\alpha$ and so $C_{Q_{\alpha+3}}(t_\alpha) \nleq Q_\alpha$. Thus $s \geq 7$.

It suffices to show that there exists a non-regular path of length 7. Let $\tilde{\gamma} = (\beta,\alpha+2,\ldots,\alpha',\alpha'+1,\alpha'+2)$ be of length 6 and pick $\alpha'+3 \in \Delta(\alpha'+2) \setminus \{\alpha'+1\}$. Since $\tilde{\gamma}$ is regular, we get $|G_{\tilde{\gamma}}/O_3(G_{\tilde{\gamma}})| = 4$ and $|O_3(G_{\tilde{\gamma}})| = 3^5$. On the other hand, by (13.11) $G_{\alpha+4}^{(3)} = V_{\alpha+4}^{(2)}$ and by (13.10) $|V_{\alpha+4}^{(2)}| = 3^5$. It follows that $V_{\alpha+4}^{(2)} = O_3(G_{\tilde{\gamma}})$.

Assume now that $(\beta,\ldots,\alpha'+3)$ is regular, i.e. $V_{\alpha+4}^{(2)} \cap Q_{\alpha'+2} \nleq Q_{\alpha'+3}$. Now by (13.5)(d) $[V_{\alpha+4}^{(2)} \cap Q_{\alpha'+2}, Z_{\alpha'+3}] = Z_{\alpha'+2}$. But $V_{\alpha+4}^{(2)}$ and $Z_{\alpha'+3}$ are in $W_{\alpha'}$ and by (13.10)(a) $W'_{\alpha'} = Z_{\alpha'}$. This implies $Z_{\alpha'} = Z_{\alpha'+2}$ which is absurd. □

<u>Remark</u>. Unlike all the other cases we have considered, it is not possible to determine the parameter r "universally", this is, the value of r depends on the realization. Indeed r may not even be finite.

Let G be the amalgamated product of G_α and G_β over B so that Γ is a tree. Then the fixed points T of K are a "line", and $|G_T| = 3|K|$. Let $n \in \mathbb{N}$, we claim that there exists a regular path of length greater than n.

Let $\tilde{\gamma}$ be any subpath of T of length n, $\tilde{\gamma} = (\alpha_0,\ldots,\alpha_n)$. It is easy to see that there exists $\delta_n \in \Gamma$ with the properties $G_T \nleq G_{\delta_n}$ and $d(\delta_n,\alpha_n) < d(\delta_n,\alpha_{n-1})$, furthermore choose δ_n with $d(\delta_n,\alpha_n)$ minimal. A similar vertex δ_0 exists corresponding to α_0. Now let γ^* be the path from δ_0 to δ_n. Then γ^* is regular and $|\gamma^*| > n$.

However, if we define

$$r_T = \max\{|\gamma| \mid \gamma \subseteq T, \ \gamma \text{ is regular}\},$$

then $r_T = 12$ and is independent of the realization. This verification is left to the reader.

14. The exchange condition.

In chapters 10 - 13 we determined some of the local structure of groups with a weak (B,N)-pair of rank 2. The question of the exact isomorphism type of the groups G_α and G_β will be subject of this and the next chapter, which finally will lead to a proof of Theorem A.

We have seen in (I. 4.4) that as long as we are interested only in the structure of G_α and G_β we may assume without loss that Γ is a tree and G is the amalgamated product of G_α and G_β. Our plan of action is to construct a normal subgroup Y of G with $Y \cap G_\alpha = Y \cap G_\beta = 1$ and with G/Y a known finite group. We will then be able to identify the images of G_α and G_β in G/Y and give their isomorphism type and embedding properties. We will not be able to do this in all cases, but rather only when $r = s-1$ and G is not of type ${}^2F_4(2)$ (for notation see chapters 3 and 4).

The main purpose of this chapter is to deduce the exchange condition for an apartment from the uniqueness condition and to prove Theorem D.

(14.0) Hypothesis. In this chapter Hypothesis B' holds and $r = s-1$. Furthermore, Γ is a tree and G is the amalgamated product of G_α and G_β over B.

We will need information about the structure of G_α and G_β determined in chapters 10 - 13. According to the value of r we divide these results into four major cases:

case	r	lemma	example
I	3	10.5	$L_3(q)$
II (a)	4	10.5	$Sp_4(q)$, q even
(b)	4	11.7 $(q_\alpha = q_\beta)$	$PSp_4(q)$, q odd
(c)	4	11.7 $(q_\alpha = q_\beta^2)$	$U_4(q)$
(d)	4	11.8	$U_5(q)$
III(a)	6	10.6	$G_2(q)$, $q \equiv 0(3)$
(b)	6	10.10 $(q_\alpha = q_\beta)$	$G_2(q)$,
(c)	6	10.10 $(q_\alpha \neq q_\beta)$	${}^3D_4(q)$,
IV	8	12.9	${}^2F_4(q)$

In the table the quoted lemmata describe some of the structure of G_α and G_β. In the last column we give an example for a group having parabolic subgroups with that structure.

Remark. (1) In all cases the stabilizers of paths of length s are p'-groups.

(2) All other cases described in chapters 10 - 13 fail to satisfy $r = s-1$; see (10.9), (12.9), (13.4), and (13.12) (and the remark there).

(3) In case III(c) $q_\alpha < q_\beta$; this is (10.7)(b).

We will refer to the above cases (and the lemmata mentioned there) by quoting I, II(a), etc.

(14.1) Let $\tilde{\gamma} = (\delta_o, \ldots, \delta_r)$ be a path of length r with $\{\delta_o, \delta_1\} = \{\alpha, \beta\}$. Then the following hold:

(a) The stabilizers of paths of length s are p'-groups.

(b) $B = Q_{\delta_o} G_{\tilde{\gamma}}$, $G_{\tilde{\gamma}} \cap Q_{\delta_o} = 1$.

(c) $G_{\tilde{\gamma}} = O_p(G_{\tilde{\gamma}})H$, where H is a B-conjugate of K.

Proof: As we have seen above (a) holds, and (b) and (c) are easy consequences of (a) and the transitivity of G_{δ_o} on paths of length r beginning at δ_o.

(14.2) Let $\tilde{\gamma}$ be as in (14.1). Then there exists $t \in N_{L_{\delta_o}}(K) \setminus B$ so that $t^2 \in K_{\delta_o}$ and $[t, K \cap G_{\delta_o}^{(1)}] = 1$. Moreover, if $p = 2$, then t can be chosen to be conjugate in L_{δ_o} to an element in $\Omega_1(O_p(G_{\tilde{\gamma}}))$.

Proof: Set $\bar{G}_{\delta_o} = G_{\delta_o}/Q_{\delta_o}$. In view of (14.1)(c) we may assume that $K \leq G_{\tilde{\gamma}}$. By the structure of $\bar{L}_{\delta_o}$ there exists a 2-element $\bar{t} \in N_{\bar{L}_{\delta_o}}(\bar{K}) \setminus \bar{B}$ so that $\bar{t}^2 \in \bar{K}_{\delta_o}$ and $[\bar{t}, \overline{K \cap G_{\delta_o}^{(1)}}] = 1$. Thus, if $p \neq 2$, we are done.

Assume now that $p = 2$. Then K has odd order and $\bar{t}$ is an involution. By (14.1) there are involutions in $L_{\delta_o} \setminus Q_{\delta_o}$, and since all involutions in $\bar{L}_{\delta_o}$ are conjugate, there is an involution t in the coset $\bar{t}$ which is conjugate to an element in $\Omega_1(O_2(G_{\tilde{\gamma}}))$.

If $[\bar{t}, \bar{K}] \neq 1$, then an easy argument applied to $\langle t \rangle K Q_{\delta_o}$ shows that t can be chosen to normalize K, and we are done.

If $[\bar{t}, \bar{K}] = 1$, then the structure of $\bar{G}_{\delta_o}$ yields $K \leq G_{\delta_o}^{(1)}$. On the other hand, we may assume that $K \neq 1$, since otherwise t has the desired properties. Hence, we are in case II(c) or (d) and $\delta_o = \beta$, or in case III(c) and $\delta_o = \alpha$.

In the first case $r = 4$, $O_2(G_{\tilde{\gamma}}) = Z_{\delta_2}$ and $[K, Z_{\delta_2}] = 1$. Thus, we may choose t in $C_{L_{\delta_o}}(K)$. In the second case $r = 6$, $O_2(G_{\tilde{\gamma}}) = Z_{\delta_3}$ and $[K, Z_{\delta_3}] = 1$, and the claim follows as above.

<u>Notation</u>. In view of (14.2) there are elements $t_\delta \in N_{L_\delta}(K) \setminus B$, $\delta = \alpha, \beta$, so that $t_\delta^2 \in K_\delta$ and $[K \cap G_\delta^{(1)}, t_\delta] = 1$. If $p = 2$, we choose t_δ so that it has the additional property described in (14.2).

Set $N = \langle t_\alpha, t_\beta \rangle K$ and $W = N/K$. Then K is a Cartan subgroup with Weyl group W (for definitions see chapter 3). Let T be the associated apartment. We label the vertices of T by $\mathbf{Z}$ with $\alpha = o$, $\beta = 1$ and $i+1 \in \Delta(i)$; i.e. $L_\alpha = L_o$, $Q_\alpha = Q_o$, $Z_\alpha = Z_o$, etc.

For any two different vertices $\delta, \lambda \in \Gamma$ let $\gamma(\delta,\lambda)$ be the unique path in Γ from δ to λ. For $i \in T$ we choose $t_i \in N \cap L_i$ so that $t_o = t_\alpha$, $t_1 = t_\beta$, $t_i \in t_o^N \cap L_i$, if $i \equiv 0(2)$, and $t_i \in t_1^N \cap L_i$, if $i \equiv 1(2)$. Furthermore we define:

$$R_i^- = O_p(G_{\gamma(i,i-r)}),$$
$$R_i^+ = O_p(G_{\gamma(i,i+r)}),$$
$$R_i = \langle R_i^-, t_i \rangle.$$

Note that by (14.1) $G_{\gamma(i,i\pm r)} = R_i^{\mp}K$, $R_i^{\mp}Q_i \in Syl_p(L_i)$, and $Q_i \cap R_i^{\pm} = 1$. Note further that $(R_i^-)^{t_i} = R_i^+$ and that t_i is a reflection on T about i.

We say that T fulfils the <u>exchange condition at</u> i, if for any $1 \neq x \in R_i^+K$ there exists $y \in R_i^-K$ so that $(i-r)^{xy} = i+r$. Note that we may assume without loss in the preceding definition that $x \in R_i^+$ and $y \in R_i^-$, since K stabilizes T.

(14.3)(a) $R_iQ_i = \langle R_i^-, R_i^+\rangle Q_i = L_i$.

(b) $K_i \leq \langle R_i^-, R_i^+\rangle$.

(c) $|R_i/\langle R_i^-, R_i^+\rangle| \leq 2$.

(d) K normalizes R_i.

Proof: By (5.1)(d) we have $L_i = \langle R_i^-, R_i^+\rangle Q_i$, and (a) follows. Since K normalizes R_i^- and R_i^+ we get (b), and (c) follows. Moreover, $[t_i, K] \leq K \cap L_i = K_i$ and so (d) holds.

(14.4) Suppose that case II(d) holds. Then $K \cap G_1^{(1)} \neq 1$ for every $i \in T$ and $[Z_o, K \cap G_o^{(1)}] \neq 1$.

Proof: We apply (11.8). Set $Y = C_K(Z_1)$, $E = Q_o/(Q_o \cap Q_1)Z_o$ and $X = C_K(E)$. Note that $K_1 \leq Y$ and that K/Y is cyclic of order $(q_1-1)/\varepsilon$, $\varepsilon = (q_1-1,2)$. It follows that $K = K_oY$ and $|K_o \cap K_1| \,\big|\, q_1+1$. This shows that K is abelian.

Note further that K_o operates elementwise fixed-point-freely and transitively on E. Hence, we get $K = K_o \times X$ and by (5.1) $X = K \cap G_1^{(1)}$.

Assume that $X = 1$. Then $K_1 \leq K_o$ and $|K_1| \,\big|\, q_1+1$. It follows that $L_1/Q_1 \simeq U_3(2)$. But it is easy to see that $U_3(2)$ cannot operate irreducibly on the module Q_1/Z_1 of order 2^6. Thus, we have $X \neq 1$, and it is easy to check that $K \cap G_o^{(1)} \neq 1$, too.

Assume now that $[Z_o, K \cap G_o^{(1)}] = 1$. Then X operates on Z_o as a subgroup of K_o. Since X centralizes Z_1 we get $|X| \,\big|\, q_1+1$. We conclude from $K = K_o \times X$ that $(q_1-1) \,\big|\, |K_o \cap K_1|$. On the other hand, as we have seen above, $|K_o \cap K_1| \,\big|\, q_1+1$ and so $q_1 = 2$ or 3.

Set $L = \langle Z_0, Z_2 \rangle$. By (5.4)(d) we have $L/O_p(L) \simeq SL_2(q_1)$. Note that $K \cap G_o^{(1)}$ centralizes L and thus a subgroup of order q_1^5 in Q_1. For $q_1 = 2$ and 3 it is easy to check that this is impossible.

(14.5)(a) $[R_i, K \cap G_i^{(1)}] = 1$.

(b) $C_K(\bar{R_i}) = K \cap G_i^{(1)}$.

(c) _If_ $K \cap G_i^{(1)} \neq 1$, _then_ $O_p(R_i) \leq G_i^{(2)}$.

Proof: Set $\overline{G_i} = G_i/Q_i$. Then $\overline{\bar{R_i}}$ is a Sylow p-subgroup of $\overline{L_i}$ and (5.1) yields $C_K(\bar{R_i}) \leq K \cap G_i^{(1)}$. On the other hand, $\bar{R_i} \cap G_i^{(1)} = 1$ by (14.1) and so $[\bar{R_i}, K \cap G_i^{(1)}] = 1$, and (b) holds. Assertion (a) follows from (b) and (14.3)(c).

Assume now that $K \cap G_i^{(1)} \neq 1$ and that $O_p(R_i) \nleq G_i^{(2)}$. Then $O_p(R_i) \nleq Q_{i+1}$, since R_i is transitive on $\Delta(i)$. If $[K \cap G_i^{(1)}, L_{i+1}] \leq Q_{i+1}$, we get a contradiction to (3.2)(c). Thus, by (5.1), (5.3)(e), and the operation of K on $O_p(R_i)/O_p(R_i) \cap Q_{i+1}$ we are in case II(d) and $i \in \alpha^G$. But then the structure of Q_i together with (a) implies $[Z_i, K \cap G_i^{(1)}] = 1$, a contradiction to (14.4). □

(14.6) _Suppose that_ $r = 4$ _and_ $K \cap G_i^{(1)} = 1$. _Then one of the following cases holds:_

(a) _Case_ II(a) _for_ $q_o = q_1 = 2$.

(b) _Case_ II(b) _for_ $q_o = q_1 = 3$ _and_ $i \in \alpha^G$.

(c) _Case_ II(c) _for_ $q_1 = 2$ or 3 _and_ $i \in \alpha^G$.

Proof: By (14.4) we are in one of the cases II(a) - (c). If we are in case II(a), it is easy to verify that $K = K_o \times K_1$. Hence, $K \cap G_i^{(1)} = 1$, if and only if $K_o = K_1 = 1$.

In cases II(b), (c) we have $[Z_1, K_1] = 1$ and $Z_{-1}Q_1 \in Syl_p(L_1)$. Moreover, $q_o > q_1$, if q_1 is even. It follows that $K_o \cap K_1$ centralizes Z_1 and Z_{-1}, and (5.1) implies $K_o \cap K_1 \leq K_1 \cap G_\beta^{(1)}$.

If q_1 is even, then $K_1 \cap G_\beta^{(1)} = 1$ and so $K = K_o \times K_1$. Hence, $K \cap G_i^{(1)} = 1$ if and only if $i \in \alpha^G$ and $q_1 = 2$, and (c) holds.

If q_1 is odd, then $|K_o \cap K_1| = 1$ or 2 and $|K_1| \geq 2$. If $K_o \cap K_1 = 1$, then $K \cap G_i^{(1)} = 1$ if and only if $i \in \alpha^G$ and $K_o = 1$, and (b) holds. If $|K_o \cap K_1| = 2$, then $K \cap G_i^{(1)} = 1$ if and only if $i \in \alpha^G$ and $K_1 \leq K_o$, and (c) holds.

(14.7) Suppose that $r > 4$ and $j = i-r/2$. Then $[K_j, R_i^-] = 1$.

Proof: Assume first that we are in case III. Then $r = 6$ and either $R_i^- = Z(L_j)$ or $R_i^- Z_j = G_j^{(2)}$ and $j \in \alpha^G$. In the first case the claim is obvious. In the second case we have $[R_i^-, L_j] = Z_j$, and it follows that $[R_i^-, K_j] \leq R_i^- \cap Z_j = 1$.

Assume now that we are in case IV. Then either $R_i^- V_j = G_j^{(2)}$ and $j \in \beta^G$ or $R_i^- Z_j = G_j^{(3)}$ and $j \in \alpha^G$. In the first case $R_i^- \cap V_j = Z_j$ and $[K_j, Z_j] = 1$, and in the second case $R_i^- \cap Z_j = 1$. Thus, we have in both cases $[R_i^-, K_j] = 1$. □

(14.8) Suppose that $r = 4$ or that case III(a) holds. Let $1 \neq x \in \Omega_1(R_i^-)$, then $x^{L_i} \cap xZ(L_i) = \{x\}$.

Proof: Set $\bar{L}_i = L_i/Z(L_i)$. It suffices to show that $C_{\bar{Q}_i}(x) = \overline{C_{Q_i}(x)}$. In each of the cases II(a) - (d) we have $C_{Q_i}(x) = Q_i \cap Q_{i-1}$, and the operation of L_i on Q_i shows the above equality. In case III(a) we have $C_{Q_i}(x) = Z_{i-2}$ and the claim follows as above.

(14.9) Suppose that $r \geq 4$ and $K \cap G_i^{(1)} \neq 1$. Then one of the following holds for $\hat{R}_i = \langle R_i^-, R_i^+ \rangle$:

(a) $O_p(R_i) = 1$.

(b) $O_p(\hat{R}_i) = 1$ and $r = 6$.

(c) $R_i = \hat{R}_i$, $r = 6$, $R_i/O_p(R_i) \simeq SL_2(q_i)$, q_i even, and $O_p(R_i)$ is a natural $SL_2(q_i)$-module.

Proof: We may assume that $O_p(R_i) \neq 1$. From (14.5) we get $[R_i, K \cap G_i^{(1)}] = 1$ and $O_p(R_i) \leq G_i^{(2)}$.

Suppose that $r = 4$. Then we are in case II, and $G_i^{(2)} \neq 1$ yields $G_i^{(2)} = Z(L_i)$. It follows that $O_p(R_i) \leq Z(L_i)$. Note that $Z(L_i) \cap R_i^- = 1$. Hence, [10,I 17.4] implies $Z(L_i) \cap [\hat{R}_i, \hat{R}_i] = 1$. Assume that $O_p(\hat{R}_i) \neq 1$. Then $[\hat{R}_i, K_i] \neq \hat{R}_i$, and we have $K_i \leq G_i^{(1)}$. It follows that $R_i = \hat{R}_i \simeq C_2 \times SL_2(2)$, $C_3 \times SL_2(3)$, or $C_2 \times SU_3(2)$. In the first two cases $\hat{R}_i$ is generated by two conjugate subgroups of order 2 or 3, respectively, namely R_i^- and R_i^+. Hence, we have $O_p(\hat{R}_i) = 1$, a contradiction. In the third case we get $[K, \hat{R}_i] = \hat{R}_i$ but $[Z(L_i), K] = 1$, again a contradiction.

Assume now that $O_p(\hat{R}_i) = 1$. Then (14.3)(c) implies $|O_p(R_i)| = 2$; in particular $q_i = 2$ and $R_i \simeq C_2 \times SL_2(2)$ or $C_2 \times SU_3(2)$. Since $t_i \notin \hat{R}_i$ but t_i is conjugate to an element in $\Omega_1(R_i^-)$, we get a contradiction to (14.8).

Suppose that $r = 6$. Then we are in case III and $[K_j, R_i] = 1$ by (14.7). Assume first that $R_j^+ \leq O_p(R_i)$. Then $O_p(R_i)Q_j \in Syl_p(L_j)$ and $K_j \leq G_j^{(1)}$ by (5.1). Hence, we have $L_j/Q_j \simeq SL_2(3)$ since $K_j \neq 1$. It follows from (3.2)(c) that $O_p(R_i) \leq Q_{i-2}$ and from (14.1)(a) that $O_p(R_i) = R_j^+$ and $O_p(R_i) \leq Z(L_i)$. Since q_i is odd we get from (14.3)(c) $R_i = \hat{R}_i$ and then as above $R_i \simeq C_3 \times SL_2(3)$ or $C_3 \times L_2(3)$, which leads to the same contradiction as before.

Assume next that $R_j^+ \nleq O_p(R_i)$. Since $G_i^{(2)} = R_j^+ Z_i$ we get $O_p(R_i) = Z_i$, $R_i/Z_i \simeq SL_2(q_i)$, and Z_i is a natural $SL_2(q_i)$-module. In addition, there is no normal subgroup of index 2 in R_i which covers R_i/Z_i and so $R_i = \hat{R}_i$. It remains to prove $q_i \equiv 0(2)$. But this is obvious, since for odd q_i $K_i \cap G_i^{(1)} \neq 1$ and $K_i \cap G_i^{(1)}$ operates fixed-point-freely on Z_i, which contradicts (14.5).

Suppose that $r = 8$. Then we are in case IV and $q_o = q_1$ is even. Moreover, $K_j \cap G_j^{(1)} = 1$ and $[K_j, R_i] = 1$ by (14.5). Hence, (5.1) and (5.3)(e) imply $G_j \cap O_p(R_i) \leq Q_j$, and we must have $O_p(R_i) \nleq Q_{i-2}$ or $O_p(R_i) \cap Q_{i-2} \nleq Q_{i-3}$. We conclude that $K_j \leq G_{i-2}^{(1)}$ or $K_j \leq G_{i-3}^{(1)}$, and K_j centralizes t_{i-2} or t_{i-3}.

If $[K_j, t_{i-2}] = 1$, then $K_j^{t_{i-2}} = K_i = K_j$ which contradicts $[K_i, R_i] \neq 1$. If $[K_j, t_{i-3}] = 1$, then $K_j^{t_j t_{i-3}} = K_j \leq G_{i-1}^{(1)}$ which contradicts (3.2)(c). □

(14.10) T <u>fulfils the uniqueness condition.</u>

<u>Proof</u>: Let $\tilde{\gamma}$ be a path of length s contained in T and T^g, $g \in G$. Then there exists $w \in N^g$ so that $\tilde{\gamma}^{gw} = \tilde{\gamma}$ and $T^{gw} = T^g$. Clearly,

$K \leq G_\gamma^{\sim}$ and by (14.1) $K = G_\gamma^{\sim}$. It follows that $gw \in K$ and $T^{gw} = T$. □

(14.11) <u>Suppose that</u> $O_p(R_i) = 1$. <u>Then</u> T <u>fulfils the exchange condition at</u> i.

<u>Proof</u>: Choose $1 \neq x \in R_i^+$. By the transitivity of R_i^- on $\Delta(i) \setminus \{i-1\}$ there is $y \in R_i^-$ so that $(i-1)^{xy} = i+1$. Since by (14.1) $R_i \cap G_{i+1} = R_i^+K$ it follows that $(R_i^-)^{xy} = R_i^+$ and thus $(R_i^-)^{xyt_i} = R_i^-$. But now $xyt_i \in R_i^-K$ and $(i-r)^{xyt_i} = i-r$ and $(i-r)^{xy} = (i-r)^{t_i} = i+r$. □

(14.12) <u>Suppose that for any</u> $1 \neq x \in R_i^+$ <u>there exists</u> $y \in R_i^-$ <u>so that</u>

(*) $\langle x,y\rangle \simeq SL_2(2)$, $L_2(3)$, <u>or</u> $SL_2(3)$.

<u>Then</u> T <u>fulfils the exchange condition at</u> i, <u>or</u> $p = 2$ <u>and</u> $C_{Q_i}(\langle x,y\rangle) \neq 1$ <u>for at least one pair</u> (x,y).

<u>Proof</u>: Set $L = \langle x,y\rangle$. We may assume without loss that x is conjugate to y in L. Then the following relation holds:

(**) $x^y = y^{x^{-1}}$.

Set $t = xyx$ and $h = tt_i'$ for $t_i' \in t_iK_i$. Then (**) implies $x^t = y$ and $y^t = x$. Moreover, $x^h \in R_i^+$ and $y^h \in R_i^-$, so h normalizes $R_i^-Q_i$ and $R_i^+Q_i$, and the structure of L_i yields $h \in KQ_i$. Thus, we can choose t_i' so that $h \in Q_i$. It follows that $[x,h] \in R_i^+ \cap Q_i = 1$ and similarly $[y,h] = 1$; i.e. $h \in C_{Q_i}(L)$.

We may assume now that $h = 1$ or $p \neq 2$. If $h = 1$, then

$$(i-r)^t = (i-r)^{t_i'} = i+r = (i-r)^{xy}$$

and T fulfils the exchange condition.

If $p \neq 2$ and $h \neq 1$, then h is an element of odd order; in particular $[t,t_i'] \neq 1$. But h centralizes L and we get $t^h = t^{t_i'} = t$, a contradiction. □

(14.13) <u>Suppose that</u> $r = 3$, <u>then</u> T <u>fulfils the exchange condition</u>.

<u>Proof</u>: We are in case I. Hence, $L_i/Q_i \simeq SL_2(q_i)$ and Q_i is a natural $SL_2(q_i)$-module.

If q_i is odd, then $K_i \cap G_i^{(1)}$ operates fixed-point-freely on Q_i, and (14.5) implies $O_p(R_i) = 1$. Now the assertion follows from (14.11).

Assume now that q_i is even. Then for every $1 \neq x \in R_i^+$ there is $y \in R_i^-$ so that $\langle x,y\rangle Q_i/Q_i \simeq SL_2(2)$. Since the elements of order 3 operate fixed-point-freely on Q_i, we get $\langle x,y\rangle \simeq SL_2(2)$ and $C_{Q_i}(\langle x,y\rangle) = 1$. Now the assertion follows from (14.12). □

(14.14) <u>Let</u> X <u>be a finite group so that</u> $X/O_3(X) \simeq PGL_2(3)$ <u>and</u> $O_3(X)$ <u>is an irreducible</u> $PGL_2(3)$-<u>module of order</u> 3^3. <u>Let</u> x_1 <u>and</u> x_2 <u>be elements of order</u> 3 <u>in</u> X <u>which are inverted in</u> X. <u>Then either</u> $\langle x_1,x_2\rangle$ <u>is a</u> 3-<u>group or</u> $\langle x_1,x_2\rangle \simeq L_2(3)$.

<u>Proof</u>: Set $L = \langle x_1,x_2\rangle$. We may assume that L is not a 3-group.

Clearly, there exists a complement H to $O_2(X)$ in X, namely $N_X(T)$, $T \in Syl_2(X')$. Pick $x_i' \in x_i O_3(X) \cap H$, $i = 1,2$, and let t be the involution in H which inverts x_i' and x_2'.

Note that x_i' does not operate quadratically on $O_3(X)$. It follows that $C_{O_3(X)}(x_i') = \langle z_i \rangle$ has order 3 and $O_3(X) = [O_3(X),x_i']\langle z_j \rangle$, $\{i,j\} = \{1,2\}$. Moreover, all elements in $x_i'[O_3(X),x_i']$ are conjugate. It is fairly easy to show that t centralizes z_1 and z_2. Thus, we get $x_i \in x_i'[O_3(X),x_i']$ and without loss $x_1 = x_1'$. Now L is conjugate to H under a suitable element in $\langle z_1 \rangle H$. □

(14.15) <u>Suppose that</u> $r = 4$. <u>Then</u> T <u>fulfils the exchange condition</u>.

<u>Proof</u>: We are in case II. If $O_p(R_i) = 1$, then by (14.11) T fulfils the exchange condition at i. Thus, we may assume that $O_p(R_i) \neq 1$. Hence, (14.9) implies that $K \cap G_i^{(1)} = 1$. We now apply (14.6).

If (14.6)(a) holds, then $R_i \simeq D_{12}$ which contradicts (14.8), since $\langle t_i \rangle$ is conjugate to $\bar{R}_i$. If (14.6)(b) holds, then $KL_i/Q_i \simeq PGL_2(3)$; in particular, $q_i \equiv 1(2)$ and $R_i = \langle \bar{R}_i, R_i^+ \rangle$. Now (14.14) yields $O_p(R_i) = 1$, a contradiction.

Assume now that (14.6)(c) holds. Note that $O_p(R_i) = Z_i$ and that Z_i is an irreducible R_i-module; i.e. $R_i = \langle \bar{R}_i, R_i^+ \rangle$. By [10,I 17.4] there exists a complement $\tilde{R}_i$ to Q_i in L_i, and without loss $\tilde{R}_i$ contains K_i.

If $q_1 = 2$, then Q_i is an orthogonal $L_2(4)$-module (and $L_i/Q_i \simeq L_2(4)$), and there are exactly two K_i-invariant complements to Q_i in $\bar{R}_i Q_i$, namely $\bar{R}_i$ and $(\bar{R}_i)^z$, $1 \neq z \in Z_{i+1}$. It follows that $\tilde{R}_i$ is conjugate to

$\langle R_i^-, R_i^+\rangle$ and $O_p(R_i) = 1$, a contradiction.

If $q_1 = 3$, then $L_i/Q_i \simeq L_2(9)$ and $|Q_i| = 3^4$; in particular $K_i \simeq C_4$. Choose $1 \neq x \in R_i^+$. It is easy to check that there is $1 \neq y \in R_i^-$ so that $D = C_{Q_i}(x) \cap C_{Q_i}(y) \neq 1$. Set $X = \langle t,x,y\rangle Q_i$, where t is the involution in K_i. Then $X/Q_i \simeq PGL_2(3)$ and $\bar{Q}_i = Q_i/D$ is an irreducible $PGL_2(3)$-module of order 3^3. Hence, (14.12) yields $\langle x,y\rangle D/D \simeq L_2(3)$. Since $|x^{Q_i}| = 9 = |x^{\bar{Q}_i}|$, we have as in (14.8) $xD \cap x^X = \{x\}$ and thus $\langle x,y\rangle \simeq L_2(3)$. Now (14.12) implies that T fulfils the exchange condition at i. $\square$

(14.16) Let X be a dihedral group of order $2n$, $1 \neq n$ odd, and let V be a $GF(2)$-module for X. Suppose that $V = V_o \times V_o^t$ where $t \in X \setminus X'$ and X' operates fixed-point-freely and transitively on V_o. Then $|X| = 6$ or V_o and V_o^t are the only X'-submodules of order $|V_o|$ in V.

Proof: Assume that E is an X'-submodule of order $|V_o|$ in V different from V_o and V_o^t. Then there exists $x \in X'$ and $1 \neq v \in V_o$ so that $vv^{xt} \in E$. Hence, for every $d \in X'$ there exists $d' \in X'$ so that

$$vv^{xt}(vv^{xt})^d = (vv^{xt})^{d'}.$$

It follows that

$$vv^d = v^{d'} \quad \text{and} \quad vv^{d^{-1}} = v^{d'^{-1}}.$$

This implies $vv^d = v^{d'} = v^{d'^{-1}d}$ and thus $d'^2 = d$ by the fixed-point-free action of X'. Similarly, we get $vv^{d'} = v^d = v^{d^{-1}d'}$ and $d^2 = d'$. This yields $d^4 = d$ for every $d \in X'$ and $|X| = 6$.

(14.17) <u>Suppose that</u> $r = 6$. <u>Then</u> t_o <u>and</u> t_1 <u>can be chosen so that</u> T <u>fulfils the exchange condition.</u>

<u>Proof</u>: Set $\hat{R}_i = \langle R_i^-, R_i^+ \rangle$ and $j = i-r/2$. Assume that T does not fulfil the exchange condition at i. Then by (14.11) $O_p(R_i) \neq 1$ and by (14.7) $K_j \leq G_i^{(1)}$.

Suppose first that $K \cap G_i^{(1)} \neq 1$. We apply (14.9). If (14.9)(c) holds, then, as in (14.13), for every $1 \neq x \in R_i^+$ there exists $y \in R_i^-$ so that $\langle x,y \rangle \simeq SL_2(2)$, and (14.12) yields a contradiction. If (14.9)(b) holds, then $R_i \neq \hat{R}_i$ and by (14.3)(c) there is a G_i-chief factor of order 2 in Q_i. Since $K \neq 1$ we are in case III(d) with $q_o = 2$ and $i \in \beta^G$. But $K = K_i$ which contradicts $K \cap G_i^{(1)} \neq 1$.

Suppose now that $K \cap G_i^{(1)} = 1$; in particular $K_j = 1$. Hence, we have $q_j = 2$. Note that j is not conjuate to i.

Assume that $q_i > 2$. Then we are in case III(d) with $j \in \alpha^G$ and $i \in \beta^G$. Set $\bar{L}_i = L_i/Z_i$. The following facts are easy consequences of the structure of L_{i-1} and L_i.

(i) $K_i = K_{i-2}$.

(ii) $|\bar{Z}_{i-1}| = 2$, $|\bar{Q}_i| = 4 \cdot q_i^2$.

(iii) $C_{\bar{Q}_i}(x) \leq \overline{G_{i-1}^{(2)}[Z_{i+1},x]}$, $1 \neq x \in R_i^-$.

(iv) $|\bar{Z}_{i-1}^{L_i}| = q_i+1$.

(v) $L_i/Q_i \simeq SL_2(q_i)$, and $\bar{Q}_i$ is an irreducible $SL_2(q_i)$-module.

From (ii), (iv) and (v) we get $q_i > 4$, i.e. $|K_i| > 3$; and from (iii) we get that every K_i-invariant complement to Q_i in $R_i^- Q_i$ is contained in $R_i^- G_{i-1}^{(2)}$ or $(R_i^-)^z G_{i-1}^{(2)}$ for $z \in Z_{i+1} \setminus Z_i$. Let A be such a complement in $R_i^- G_{i-1}^{(2)}$. Note that $G_{i-1}^{(2)} = R_{i-4}^+ Z_{i-2}$ and $[Z_{i-2}, K_i] = 1$. It follows that $A \leq R_i^- R_{i-4}^+$. Now (14.16) implies $A = R_i^-$. We conclude that R_i^- and $(R_i^-)^z$ are the only K_i-invariant complements to Q_i in $R_i^- Q_i$. Now [10, I 17.4] applied to $\bar{L}_i$ shows that there is a complement to $\bar{Q}_i$ in $\bar{L}_i$, and as we have seen, this complement is conjugate to $\hat{R}_i Z_i / Z_i$. Hence, $O_p(\hat{R}_i) \leq Z_i$ ans as in (14.9) we get from $[\hat{R}_i, K_i] = \hat{R}_i$ that $O_p(\hat{R}_i) = 1$; i.e. $R_i = Z(L_i) \times \hat{R}_i$.

Choose $t_i' \in t_i Z(L_i) \cap \hat{R}_i$ and set $\tilde{N} = \langle t_{i-1}, t_i' \rangle K$, $\tilde{W} = \tilde{N}/K$. Let $\tilde{T}$ be the associated apartment. For $\tilde{T}$ we use the same notation as for T but decorated with ~. We want to show that $\tilde{T}$ fulfils the exchange condition. As we have seen above, the only vertices in $\tilde{T}$ which cause trouble are conjugate to our fixed i. Hence, it suffices to show that $O_p(\tilde{R}_i) = 1$.

Note that $(i-4,\dots,i+3) \subseteq T \cap \tilde{T}$, since $Z(L_i)$ fixes $\Delta^{(3)}(i)$ pointwise. If $R_i^+ = \tilde{R}_i^+$, then clearly $\tilde{R}_i = \langle t_i', R_i^+ \rangle = \hat{R}_i$ and $O_p(\tilde{R}_i) = 1$. Hence, we have $\tilde{R}_i^+ = (R_i^+)^z$ for $z \in Z_{i-1} \setminus Z_i$. On the other hand, $(i+2)^z \neq (i+2)$ and $\tilde{R}_i^+$ stabilizes $(i,i+1,\dots,(i+6)^z)$ and $(i,\dots,i+3,\dots,i+6) \subseteq \tilde{T}$. Hence, $\tilde{R}_i^+$ stabilizes $((i+6)^z,\dots,i+1,\dots,i+6)$, a path of length 10. This contradicts (14.1)(a).

Assume now that $q_0 = q_1 = 2$. Then $K = 1$ and R_i^- has order 2. Set $R_i^- = \langle r \rangle$. Suppose first that $i \in \alpha^G$. Then the structure of L_i implies $R_i \simeq D_{12}$ and $O_2(R_i) = R_{i-3}^+$. Set $R_{i-3}^+ = \langle r' \rangle$. By our choice of t_i we have that r is conjugate to rr' in L_i. On the other hand, by the structure of Q_{i-3} we have also have $R_i^- = Z_{i-3}$ and so $\langle rr' \rangle = Z_\lambda$ for some $\lambda \in \Gamma$ with $d(\lambda,i) = 3$ and $d(\lambda,i-3) \equiv 0(2)$. It follows that $i-3,\lambda \in \Delta(i-2)$ and $r, rr' \in Z_{i-2}$. But Z_{i-2} is a natural $SL_2(2)$-module for L_{i-2}/Q_{i-2} and so $\langle r' \rangle = Z_{i-3}$ or Z_{i-1} which is absurd.

Suppose next that $i \in \beta^G$. Then the structure of L_i yields $R_i \simeq D_{12}$ or D_{24}. Let d be an element of order 3 in R_i with $(i-1)^d = i+1$ and set $\Omega = \{i-2, (i-2)^d, (i-2)^{d^{-1}}\}$. Note that $r \in G_{i-2}$ and $t_i \in G_{i-2}^{d^{-1}}$, since $t_i \in Q_{i-1}^{d^{-1}}$ by our choice of t_i. It follows that $\langle r,d\rangle$ and $\langle t_i,d\rangle$ operate on Ω and

$$(i-2)^{dt_ir} = i-2 \quad \text{and} \quad (i-2)^{d(dt_ir)} = (i-2)^d.$$

Hence, we get

$$dt_ir \in Q_{i-1} \cap Q_{i+1} \cap Q_{i-1}^{d^{-1}} = G_i^{(2)} = Z_i ,$$

again by the structure of L_i. If $d = rt_i$, then $\langle r,t_i\rangle = R_i \simeq SL_2(2)$ which contradicts $O_2(R_i) \neq 1$. Thus, we have $d = rt_iz_i$, $1 \neq z_i \in Z_i$.

We now proceed as above. Set $\tilde{N} = \langle t_{i-1}, t_iz_i\rangle$ and let $\tilde{T}$ be the associated apartment. As above $\tilde{T}$ fulfils the exchange condition, if $\tilde{\bar{R}}_i = \bar{R}_i$. Assume that $\bar{R}_i \neq \tilde{\bar{R}}_i$. Since $\bar{R}_i\tilde{\bar{R}}_i \leq Q_{i-4} \cap G_i$ we get $\tilde{\bar{R}}_i \leq \bar{R}_iZ_{i-2}$ and so $\tilde{\bar{R}}_i = \langle rz\rangle$, $1 \neq z \in Z_{i-2}$. On the other hand, $\langle Z_{i-2}^{L_i}\rangle \simeq C_4 * Q_8$ and $\langle r,d\rangle\langle Z_{i-2}^{L_i}\rangle/Z_i \simeq C_2 \times \Sigma_4$, and it is easy to check in this group that $\langle rz, t_iz_i\rangle \simeq D_{24}$. But with the above argument we get for $\tilde{T}$ in place of T that $\langle \tilde{\bar{R}}_i, t_iz_i\rangle Z_i/Z_i \simeq SL_2(2)$. This contradiction completes the proof. □

(14.18) <u>Suppose that</u> $r = 8$. <u>Then</u> T <u>fulfils the exchange condition or</u> G <u>is of type</u> ${}^2F_4(2)$.

<u>Proof:</u> We are in case IV. If $q_0 = q_1 = 2$, then G is of type ${}^2F_4(2)$. Hence, we may assume that $q_0 = q_1 > 2$ and thus $K_i \neq 1$ for $i \in T$. From (14.5), (14.7) and (14.9) we get $O_p(R_i) = 1$ for $i \in T$. Now (14.11) yields the assertion. □

The next lemma sums up what we have proven so far.

(14.19) _One of the following holds:_

(a) G _is of type_ $^2F_4(2)$.

(b) T _can be chosen to fulfil the uniqueness and exchange condition._

Theorem D. _One of the following holds:_

(a) G _is of type_ $^2F_4(2)$.

(b) G _is locally isomorphic to_ $L_3(p^n)$, $PSp(p^n)$, $G_2(p^n)$, $U_4(p^n)$, $U_5(p^n)$, $^3D_4(p^n)$, _or_ $^2F_4(2^n)$.

Proof: We may assume that G is not of type $^2F_4(2)$. According to (14.19) there exists an apartment $T = T(K,W)$ which fulfils the uniqueness and exchange condition. We apply (3.6) and (3.7). Then there exists a normal subgroup D of G with

(i) $G_\alpha \cap D = G_\beta \cap D = 1$.

(ii) G/D has a (B,N)-pair of rank 2 (for our fixed B and N).

Since $B = KS$ this (B,N)-pair is split and the classification of Fong and Seitz [6] applies to G/D. This yields the assertion, since G is locally isomorphic to G/D. □

15. Proof of Theorem A.

We collect the results proven in chapters 7 - 14 to give a proof of Theorem A here. We assume throughout that G is a group with a weak (B,N)-pair of rank 2; i.e. G satisfies Hypothesis A or equivalently Hypothesis A'. In the following we use Hypothesis A' and the notation defined in chapter 3. As in chapter 14 we may assume that Γ is a tree.

Assume first that G satisfies, in addition, Hypothesis B'. Suppose that $[Z_\alpha, Z_{\alpha'}] \neq 1$, then (7.3) and (7.5) show that $b \leq 2$. Suppose that $[Z_\alpha, Z_{\alpha'}] = 1$, then (8.4) and (9.11) show that $b = 1,3,5$. This proves Theorem B.

Now suppose that it is not the case that stabilizers of paths of length s are p'-groups. Then (10.5), (10.6), (10.9), (10.10), (11.7), (12.9), (13.4), and (13.12) prove Theorem C.

Theorem D was already proven in chapter 14.

We now turn to Theorem A whose proof is not quite lemmata quoting. For the convenience of the reader we repeat its statement here.

Theorem A. Suppose that G is a group with a weak (B,N)-pair of rank 2. Then one of the following holds:

(a) G is locally isomorphic to X where $X_o \leq X \leq \mathrm{Aut}(X_o)$ and $X_o \simeq L_3(p^n)$, $PSp_4(p^n)$, $U_4(p^n)$, $U_5(p^n)$, $G_2(p^n)$, ${}^3D_4(p^n)$, ${}^2F_4(2^n)$.

(b) G is parabolic isomorphic to $G_2(2)'$, J_2, $\mathrm{Aut}(J_2)$, M_{12}, or $\mathrm{Aut}(M_{12})$.

(c) G is of type ${}^2F_4(2)'$, ${}^2F_4(2)$, or F_3.

Notation.

$B_o = (B \cap G^*_\alpha)(B \cap G^*_\beta)O_p(B)$,

$L_\delta = G^*_\delta B_o$, $\delta = \alpha,\beta$,

$M = \langle L_\alpha, L_\beta \rangle$,

$r = r(\Gamma,M)$,

$s = s(\Gamma,M)$,

K is a complement to $O_p(B_o)$ in B_o.

(15.1)(a) B_o is p-closed.

(b) M is normal in G.

(c) M satisfies Hypothesis B' with respect to Γ.

(d) If $K = 1$, then $G = M$.

Proof: (a) follows from the fact that $G^*_\delta \cap B$ is p-closed (see also chapter 4), (b) follows from (3.2), and (c) is a consequence of (a).

Assume that $K = 1$. Then Theorems C and D imply that $L_\delta/Q_\delta \simeq SL_2(2)$ or $Sz(2)$, $\delta = \alpha,\beta$. Let T be a Sylow q-subgroup in B for $q \neq p$. Then $N_{G_\delta}(T)$ is transitive on $\Delta(\delta)$ for $\delta = \alpha,\beta$ and (3.2)(c) yields $T = 1$. It follows that $B = B_o$ and $G = M$. □

(15.2) Suppose that $r = s-1$. Then case (a) of Theorem A holds, or G is of type ${}^2F_4(2)$.

Proof: By (15.1)(c) M satisfies Hypothesis (14.0). Thus, we can apply Theorem D to M. If B is p-closed, then G also satisfies Hypothesis (14.0), and the assertion follows from Theorem D. Thus, we may assume that B is not p-closed; in particular, by (15.1) $B \neq B_0$ and $K \neq 1$, and M is not of type ${}^2F_4(2)$.

By (14.2) K is a Cartan subgroup and by (14.19) there exists an apartment $T = T(K)$ which fulfils the uniqueness and exchange condition. As in chapter 14 we choose T to contain (α,β) and label its vertices by $\mathbf{Z}$ so that $\alpha = o$ and $\beta = 1$. Let $\tilde{\gamma}$ be a path of length s in T. By a Frattini argument we have $G = G_{\tilde{\gamma}}M$.

According to (3.6) the set $A = \{T^m \mid m \in M\}$ defines an equivalence relation $\approx$ on Γ so that $\Gamma/\approx$ is a generalized r-gon. Moreover, by Theorem D M contains a normal subgroup D so that M/D is a Chevalley group of rank 2, and it is clear from the proof of Theorem D that D is the kernel of the action of M on Γ.

Suppose that $G_{\tilde{\gamma}} = G_T$. Then G normalizes A and operates on $\Gamma/\approx$. Let D_o be the kernel of the action of G on $\Gamma/\approx$. Then $G/D_o \leq Aut(MD_o/D_o)$ and the assertion follows.

Suppose that $G_{\tilde{\gamma}} \neq G_T$. We choose notation as in chapter 14 so that $N = \langle t_o,t_1\rangle K$, $t_i \in N \cap G_i$ is conjugate to t_o or t_1, respectively, and $\tilde{\gamma} = (i,\ldots,i+s)$, $i \in T$. Note that $N = \langle t_i,t_{i+1}\rangle K$.

We may assume without loss that $G_{\tilde{\gamma}} \nleq G_{i-1}$. Let $n \in T$, $n \geq i+s$, be minimal so that $G_{\tilde{\gamma}} \nleq G_{n+1}$. Note that $G_{\tilde{\gamma}} \cap M = K$ and so K is normal in $G_{\tilde{\gamma}}$. Hence, $G_{\tilde{\gamma}}$ operates on the fixed points of K on $\Delta(n)$ (resp. $\Delta(i)$). Now (5.1) implies that $K \leq M_n^{(1)} \cap M_i^{(1)}$. We conclude from Theorem D and the fact that $K \neq 1$ that one of the following holds:

(i) M is locally isomorphic to ${}^3D_4(2)$ and $i,n \in \alpha^G$.

(ii) M is locally isomorphic to $PSp_4(3)$ and $i,n \in \beta^G$.

(iii) M is locally isomorphic to $U_4(2)$ and $i,n \in \beta^G$.

Set $B_1 = O^p(O^{p'}(B))$. Then B_1 is non-trival, since B is not p-closed. In particular, there exists a non-trivial Sylow q-subgroup $\tilde{S}$ in B_1, $q \neq p$.

In case (i) $M_i/M_i^{(1)} \simeq SL_2(2)$ and $M_{i+1}/O_2(M_{i+1}) \simeq SL_2(8)$, and the structure of $Aut(SL_2(8))$ shows that $[\tilde{S},K] \leq O_p(B)$, and it follows that $\tilde{S} \leq M_i^{(1)} \cap M_{i+1}^{(1)}$. Now an easy application of (3.2)(c) to $N_{M_j}(\tilde{S})$, $j = i,i+1$, yields a contradiction.

In case (ii) $M_i/O_3(M_i) \simeq SL_2(3)$ and $M_{i+1}/O_3(M_{i+1}) \simeq PGL_2(3)$, and as above we get $\tilde{S} \leq M_i^{(1)} \cap M_{i+1}^{(1)}$ and a contraction to (3.2)(c).

Suppose now that we are in case (iii). The local structure of M is described in (11.7); from where we get:

(iv) $s = 5$ and $|Z_i| = 2$.

(v) $C_{O^{2'}(M_{i+2})}(K) = \langle Z_i,Z_{i+4}\rangle Z_{i+2}$.

(vi) $K \cap O^{2'}(M_{i+2}) = 1$.

Note that by (vi) $[K,t_{i+2}] \leq K \cap O^{2'}(M_{i+2}) = 1$. Hence, (iv) and (v) imply that $G_{\tilde{\gamma}}$ centralizes t_{i+2}. But then $G_{\tilde{\gamma}}$ stabilizes $\tilde{\gamma}^{t_{i+2}}$ and $G_{\tilde{\gamma}} \leq G_{i-1}$, a contradiction. □

(15.3) <u>Suppose that</u> $r \neq s-1$. <u>Then case</u> (b) <u>or</u> (c) <u>of Theorem</u> A <u>holds</u>.

<u>Proof</u>: We proceed as in (15.2). If B is p-closed, then assertion follows from Theorems C and D.

Suppose now that B is not p-closed; i.e. $G \neq M$ and $K \neq 1$. Then Theorems C and D imply that M is of type F_3 or parabolic isomorphic to J_2.

If M is of type F_3 then $M_\delta/Q_\delta \simeq GL_2(3)$ for every $\delta \in \Gamma$, and as in (15.2) we get a contradiction to (3.2)(c).

Assume now that M is parabolic isomorphic to J_2. We want to show that G is parabolic isomorphic to $Aut(J_2)$. For this purpose we use the same method as in (13.2), namely we prove that G_α and G_β are uniquely determined by generators and relations derived from the action of G on Γ.

From the local structure of M we get the following information (see also chapter 10):

(i) $M_\alpha/Q_\alpha \simeq C_3 \times SL_2(2)$, and the elements of order 3 in $O^{2'}(M_\alpha)$ operate fixed-point-freely on Q_α.

(ii) $M_\beta/Q_\beta \simeq SL_2(4)$, $Q_\beta \simeq D_8 * Q_8$, and Q_β/Z_β is an orthogonal $SL_2(4)$-module.

(iii) $N_{M_\alpha}(K) \simeq C_3 \times \Sigma_4$.

(iv) $K[Q_\beta,K] \simeq SL_2(3)$, and there exists $t \in N_{M_\beta}(K) \setminus B$, $t^2 = 1$.

We choose $1 \neq k \in K$; $d \in N_{M_\alpha}(K) \cap O^{2'}(M_\alpha)$, $o(d) = 3$; $1 \neq z \in Z_\beta$; $t \in N_{M_\beta}(K) \setminus B$, $t^2 = 1$; $q \in [Q_\beta,K]$, $o(q) = 4$. It follows that

(v) $\langle k,d,z,z^{dt},q \rangle = M_\alpha$,

(vi) $\langle k,t,q,q^d \rangle = M_\beta$.

Moreover, the relations between these generators are determined, since the isomorphism type of M_α and M_β is known.

With the help of (iii), (iv) and (3.2)(c) it is easy to check that $|B/B_0| = 2$ and

(vi) $N_{G_\alpha}(K) \simeq \Sigma_3 \times \Sigma_4$,

(vii) $G_\beta/Q_\beta \simeq \Sigma_5$.

Hence, we can choose an involution $x \in N_B(K)$ which commutes with d and z and inverts k; moreover, $<x,k>[Q_\beta,k] \simeq GL_2(3)$ and so the order of xq is known. It follows that the isomorphism type of G_α is uniquely determined.

It remains to determine the order of xt. If there is only one possibility for the order of xt, then G is parabolic isomorphic to $Aut(J_2)$, since $Aut(J_2)$ provides an example for such a configuration.

We set the following notation. $\bar{G}_\beta = G_\beta/Z_\beta$, and E is the preimage of $G_{\bar{Q}_\beta}(\bar{x})<\bar{x}>$ in G_β. Note that $\bar{x}$ induces a transvection in $\bar{G}_\beta$; i.e. $|Q_\beta/Q_\beta \cap E| = 2$. Note further that there is an element d' of order 3 in G_β which normalizes xQ_β. Hence, d' normalizes E. Set $Q = [Q_\beta,d']$, $C_0 = C_{Q_\beta}(d')$, and $C_1 = C_0 \cap E$. Then we get from (ii)

(viii) $Q \simeq Q_8$, $C_0 \simeq D_8$,

(ix) $E \cap Q_\beta = C_1 * Q \simeq C_4 * Q_8$.

The involution x normalizes C_0 and interchanges the two subgroups conjugate to Z_α which are in C_0. It follows that x inverts the elements in C_1.

Assume first that $[d',x] = 1$. Then the Three-Subgroup Lemma yields $[Q,x] = 1$. On the other hand, Z_α is contained in $C_1 * Q$ and so $C_1Q = Z_\alpha Q$ is centralized by x. This contradicts $[C_1,x] = C_1$.

Assume now that $[d',x] \neq 1$. Then $x = cq'$, $c \in C_E(d') \setminus Q_\beta$ and $q' \in Q$. Since q' has order 4, c has order 4, too. Hence, there are two different cyclic groups of order 4 in $C_E(d')$, namely $<c>$ and C_1.

Moreover, x and thus c inverts the elements in C_1. It follows that $C_E(d') \simeq Q_8$.

We now determine the order of xt. Since $x \notin M$ and $t \in M \setminus B$ we have $o(xt) = 6$ modulo Q_β. In addition, k is conjugate to d' in G_β and $(xt)^3 Q_\beta$ is conjugate to xQ_β. It follows that $(xt)^3 \in C_{E^g}(k) \simeq Q_8$ for suitable $g \in G_\beta$. Hence, xt has order 12. □

The proof of Theorem A is now a consequence of (15.2) and (15.3).

16. Problems, exercises and generalizations.

We want to use this last chapter to present a number of possible generalizations of Theorem A and point out some remarkable behaviour occurring in certain of the amalgams we have considered.

We begin by considering Hypothesis A (ii), namely $C_{P_i}(O_p(P_i)) \leq O_p(P_i)$. Suppose that Hypothesis A (i), (iii) hold and further that

$$\text{(ii)}' \quad C_{P_i}(O_p(P_i)) \not\leq O_p(P_i), \quad i = 1,2.$$

Examples of such configurations are easy to construct. Simply let

$$M_i \in \{L_2(p^{n_i}),\ U_3(p^{n_i}),\ Sz(2^{n_i}),\ Ree(3^{n_i})\}$$

for $i = 1,2$ with both M_i defined over fields of the same characteristic. Set $G = M_1 \times M_2$, let $S \in Syl_p(G)$ and set $P_i = M_i N_G(S)$, $i = 1,2$. Then G is an amalgam of P_1 and P_2 over $B = N_G(S)$, and $O_p(P_i) \in Syl_p(M_{3-i})$. Obviously, these examples can be decorated with automorphisms of the groups M_i. We will call any such configuration defined above a <u>direct product amalgam</u>.

<u>Exercise</u> 1. Show that if Hypothesis A (i) and (iii) and, in addition, (ii)' hold, then either G is parabolic isomorphic to a direct product amalgam or $p = 2$ or 3, and $n_i = 1$, $i = 1,2$. Describe P_i in the latter case.

Under the hypothesis of Exercise 1, and after excluding the "small" cases, one can even show that G is locally isomorphic to a direct product amalgam. The key is to notice that $r = s-1 = 2$ and then to use the techniques of chapter 14.

The next case to consider is that in which Hypothesis A (i), (iii) hold and further that

$$\text{(ii)}'' \quad C_{P_i}(O_p(P_i)) \not\leq O_p(P_i) \quad \text{for exactly one} \quad i \in \{1,2\}.$$

Examples of such configurations are again plentiful although somewhat difficult to describe. Let

$$M_i \in \{SL_2(p^{n_i}),\ L_2(p^{n_i}),\ U_3(p^{n_i}),\ SU_3(p^{n_i}),\ Sz(2^{n_i}),\ Ree(3^{n_i})\}$$

for $i = 1,2$ with the additional property that $Z(M_1) = 1$. Now let $G = M_1 \int M_2$, the wreath product of M_1 by M_2 with respect to the natural doubly transitive permutation representation of M_2 of degree $p^{n_2}+1$, $p^{2n_2}+1$, or $p^{3n_2}+1$, respectively. Then G is the semidirect product of a normal subgroup N of G by M_2, where N is the direct product of n copies of M_1, n the degree of the permutation representation. Let $S_2 \in Syl_p(M_2)$, $S_1 \in Syl_p(N)$, so that $M_2 \leq N_G(S_1)$, and set $N_2 = N_{M_2}(S_2)$, $N_1 = N_N(S_1)$, $B = N_1N_2$. Then we may assume without loss that N_2 normalizes M_1. Let $P_1 = BM_1$ and $P_2 = BM_2$. Then G is an amalgam of P_1 and P_2 over B, furthermore $C_{P_1}(O_p(P_1)) \not\leq O_p(P_1)$, $C_{P_2}(O_p(P_2)) = C_{P_2}(S_1) \leq S_1 \leq O_p(P_2)$. These examples can again be decorated with appropriate automorphisms as above. We will call any such configuration defined above a <u>wreath product amalgam</u>.

<u>Exercise</u> 2. Show that if Hypothsis A (i), (iii) and, in addition, (ii)" hold, then either G is parabolic isomorphic to a wreath product amalgam or $p = 2$ or 3. Describe P_i in the latter case.

There is an obvious "extra case" of Hypothesis A which we would like to point out. For any odd prime p we have mentioned in (5.2) that for a p-element $x \in SL_2(p^n)$ there exists a conjugate element x^g so that $\langle x,x^g\rangle = SL_2(p^n)$ except for the case $p^n = 9$ in which case the subgroup of largest order generated in this way is isomorphic to $SL_2(5)$.

<u>Exercise</u> 3. Carry out the proof of Theorem A under the assumption that for at least one $i \in \{1,2\}$

$$P_i^*/O_3(P_i) \simeq L_2(5) \quad \text{or} \quad SL_2(5).$$

Determine the structure of P_1 and P_2.

One interesting example of this situation lives in the simple group M^cL constructed by McLaughlin. It contains subgroups of the form $3^5GL_2(5)$ and 3^4M_{10} intersecting in a Sylow 3-normalizer of order $3^6\ 2^3$. (Note that M_{10} is an extension of $PSL_2(9)$).

One further "degeneracy" is related to the group $U_3(2)$ which has a normal subgroup isomorphic to $C_3 \times C_3$. Note that $Aut(U_3(2))/O_3(U_3(2)) \simeq GL_2(3)$. If one only assumes that for one $i \in \{1,2\}$ $P_i/O_{2,3}(P_i)$ is a subgroup of $GL_2(3)$ and $|O_{2,3}(P_i)/O_2(P_i)| \leq 3^3$, then one example of this is easy to describe. Let $G = \Sigma_8$ and $S \in Syl_2(G)$. Then there are two maximal subgroups M_1, M_2 of G containing S so that $M_1/O_2(M_1) \simeq SL_2(2)$ and $M_2/O_2(M_2) \simeq SL_2(2) \int C_2$; i.e. $M_2/O_{2,3}(M_2) \lneqq GL_2(3)$.

<u>Exercise</u> 4. Carry out the proof of Theorem A under the above assumption. Determine the type of G.

We next turn to the conclusion of Theorem A. There is an obvious discontent here. We refer, of course, to the groups being "parabolic isomorphic" to $G_2(2)'$, J_2, M_{12}, $Aut(M_{12})$, $Aut(J_2)$ and those being of "type" ${}^2F_4(2)'$, ${}^2F_4(2)$, or F_3. The ideal conclusion would replace these somewhat spongey terms with the term "locally isomorphic". Of course, in almost all applications to finite group theory these indeterminacies are perfectly acceptible.

First, in order to improve "type" to "parabolic isomorphic" an analysis of generators and relations as in chapter 13 seems called upon. The next exercise is precisely this.

<u>Exercise</u> 5. Show that a group G of type ${}^2F_4(2)'$, ${}^2F_4(2)$, or F_3 is parabolic isomorphic to ${}^2F_4(2)'$, ${}^2F_4(2)$, or F_3, respectively.

This analysis has been carried out for G of type ${}^2F_4(2)'$ and ${}^2F_4(2)$ by Fan [5] yielding the desired conclusion. For a group of type F_3 the analysis has never been carried out in full and appears to be very difficult (but see Exercise 13).

In order to make the last leap from "parabolic isomorphic" to "locally isomorphic" there is a general procedure due to Goldschmidt [7] which we now describe. Let P_1, P_2, B be any three groups and ϕ_i, $i = 1,2$, be a pair of monomorphisms from B into P_i. This data,

$$P_1 \xleftarrow{\ \phi_1\ } B \xrightarrow{\ \phi_2\ } P_2$$

uniquely determines the isomorphism type of $G(\phi_1,\phi_2)$, the amalgamated product of P_1 and P_2 over B with respect to ϕ_1 and ϕ_2. However, for a different pair of monomorphisms $(\tilde{\phi}_1,\tilde{\phi}_2)$ the associated amalgamated product $G(\tilde{\phi}_1,\tilde{\phi}_2)$ need not be isomorphic to $G(\phi_1,\phi_2)$ although the two groups are clearly parabolic isomorphic. In order to prove that parabolic isomorphism implies local isomorphism it must be shown that $G(\phi_1,\phi_2) \simeq G(\tilde{\phi}_1,\tilde{\phi}_2)$ for any two pairs of monomorphisms.

To this end let A_i^* be the subgroup of $\mathrm{Aut}(P_i)$ consisting of those automorphisms of P_i leaving $\mathrm{im}\,\phi_i$ invariant, and define

$$\phi_i^* : A_i^* \longrightarrow \mathrm{Aut}(B)$$

via $\psi\phi_i^* = \phi_i\psi\phi_i^{-1}$. Let $A_i = \mathrm{im}\,\phi_i^*$. Then by [7,2.7]:

$\mathrm{Aut}(B) = A_1A_2$ if and only if $G(\phi_1,\phi_2) \simeq G(\tilde{\phi}_1,\tilde{\phi}_2)$ for all pairs of monomorphisms.

<u>Exercise</u> 6. Show that if G is parabolic isomorphic to $G_2(2)'$, J_2, M_{12}, $Aut(M_{12})$, $Aut(J_2)$, ${}^2F_4(2)'$, ${}^2F_4(2)$, or F_3, respectively, then G is locally isomorphic to the corresponding group.

This analysis has been carried through by Goldschmidt [7] in the cases $G_2(2)'$, M_{12}, $Aut(M_{12})$, and by Fan [5] in the cases ${}^2F_4(2)'$, ${}^2F_4(2)$ yielding the desired conclusion.

A further possibility for a solution to the above exercise would follow the idea of chapters 3 and 14. Namely, construct an appropriate quotient of the associated tree and identify the induced group by quoting some theorems. However, the very individual nature of these different groups makes this approach very difficult. For a group of type F_3 one would probably need a characterization of F_3 in terms of its 3-structure, which does not seem to exist.

We next discuss "subamalgams". We will not define this term formally but rather give some examples that will make the notion clear. We will assume throughout this part of the discussion that Hypothesis B holds and Γ is a tree.

Suppose first that $r = s-1$ and case (a) of Theorem A holds. According to (14.2) there exists a Cartan subgroup K with associated apartment T. Let $\lambda_i = (\alpha_{(i-1)r},\dots,\alpha_{ir})$, $i = 1,2,3$, be three paths of length r in T, and set $M_i = \langle G_{\lambda_i}, G_{\lambda_{i+1}}\rangle$, $i = 1,2$, and $X = \langle M_1, M_2\rangle$. Suppose that $M_i \simeq G_{\alpha_{ir}}/O_p(G_{\alpha_{ir}})$ (this property can be deduced from the local structure of G, if T is chosen as in chapter 14).

Exercise 7. $X/Z(X)$ is a direct product amalgam.

There are several such subamalgams which can be reaped from the classical amalgams. Suppose that G is locally isomorphic to $G_2(p^n)$, $p \neq 3$. Let $\alpha_0 \in \Gamma$ with $Z(O^{p'}(G_{\alpha_0})) = 1$ and let $\lambda_1 = (\alpha_0,\ldots,\alpha_6)$ and $\lambda_2 = (\alpha_2,\ldots,\alpha_8)$ be two paths of length 6. Set $M_1 = \langle Z_{\alpha_0}, Z_{\alpha_4}\rangle G_{\lambda_1}$, $M_2 = \langle Z_{\alpha_2}, Z_{\alpha_6}\rangle G_{\lambda_2}$ and $X = \langle M_1, M_2\rangle$.

Exercise 8. $X/Z(X)$ is locally isomorphic to $L_3(p^n)$.

Following the same pattern:

Exercise 9. Show that if G is locally isomorphic to $G_2(3^n)$, ${}^2F_4(2^n)$, or ${}^3D_4(p^n)$, respectively, then G contains a subgroup X so that $X/Z(X)$ is locally isomorphic to $L_3(3^n)$, $Sp_4(2^n)$, or $L_3(p^n)$, respectively.

We have so far discussed families of subamalgams. There are a number of "sporadic" subamalgams of some interest. The first two we discuss can be found in a group locally isomorphic to $G_2(4)$. We have proven in (10.10) that, with the notation there, Q_β/Z_β is an irreducible G_β/Q_β-module. However, in this one case $q = 4$, it is not an irreducible L_β/Q_β-module, rather a direct sum of two orthogonal L_β/Q_β-modules.

Exercise 10. Show that there is a subgroup M in L_β so that $M/O_2(M) \simeq SL_2(4)$ and $O_2(M)$ is extra special of order 2^5.

Exercise 11. Show that if G is locally isomorphic to $G_2(4)$ then G contains a subgroup X which is parabolic isomorphic to J_2.

Let G now be of type ${}^2F_4(2)$ and let $X = \langle Z_\delta / \delta \in \Gamma \rangle$.

Exercise 12. X is of type ${}^2F_4(2)'$.

The final subamalgam we describe is truely exotic. Let G be of type F_3. We refer to chapter 13 for notation. Let K be a complement to S in $G_\alpha \cap G_\beta$; i.e. $K \simeq C_2 \times C_2$. Then $C_{G_\alpha^{(3)}}(K) \simeq C_3$. Let $1 \neq g \in C_{G_\alpha^{(3)}}(K)$ and set $X = C_G(g)$.

Exercise 13. $X/\langle g \rangle$ is locally isomorphic to $G_2(3)$.

This exercise shows that any amalgam G of type F_3 must necessarily contain a subamalgam of the given form. This should aid in giving generators and relations for G_α and G_β. A more geometric description of this subamalgam can be given by noticing that the fixed points of K on Γ, say T, admit an element τ operating as a 2-translation on T, that is $d(\delta, \delta\tau) = 2$ for all $\delta \in T$. Set $Y_\beta = \langle G_\alpha^{(3)\tau^{-1}}, G_\alpha^{(3)}, G_\alpha^{(3)\tau} \rangle$ and $Y = \langle Y_\beta, Y_\beta^\tau \rangle$. Then $X = Y$.

Additional subamalgams can be constructed by taking any apartment T and $1 \neq k \in G_T$. Set $X = C_G(k)$. Then $X/Z(X)$ is in general a direct product amalgam or a wreath product amalgam although, as we have seen above, there are exceptions.

Finally we comment on generalizations to higher rank amalgams. To avoid extensive notation we restrict ourselves to generalizations of Hypothesis B.

Consider for $n \in \mathbb{N}$, $n \geq 2$, the following

Hypothesis B^n. Let p be a prime and G an amalgam of the finite groups $P_1,\ldots,P_n$ over B. Suppose that for each $i \in \{1,\ldots,n\}$:

(i) B is the normalizer of a Sylow p-subgroup in P_i.

(ii) $O^{p'}(P_i/O_p(P_i))$ is a rank 1 Chevalley group over a field of characteristic p.

(iii) $C_{P_i}(O_p(P_i)) \leq O_p(P_i)$.

Note that Hypothesis B is in this notation Hypothesis B^2 (with the addition of D_{10} and Ree(3)' as allowable sections). The case $n \geq 3$ has not yet been touched in that generality. Nevertheless, there are some nice and interesting results under the following

Hypothesis B^n_+. Hypothesis B^n holds and for $P_{ij} = \langle P_i, P_j \rangle$, $i \neq j$ and $i,j \in \{1,\ldots,n\}$:

(iv) $O^{p'}(P_{ij}/O_p(P_{ij}))$ is a rank 2 Chevalley group defined over a field of characteristic p.

(v) B is the normalizer of a Sylow p-subgroup in P_{ij}.

There is a particularly easy way to visualize Hypothesis B^n_+. Draw a graph with n vertices $\{1,\ldots,n\}$ and labeled edges so that two vertices i and j are not joined, if $O^{p'}(P_{ij}/O_p(P_{ij}))$ is a direct product amalgam (i.e. $SL_2(q_i) \times SL_2(q_j)$, etc.), and i, j are joined by an edge labeled with U, if $O^{p'}(P_{ij}/O_p(P_{ij}))$ is isomorphic to U.

The first result in this direction is a beautiful paper by R. Niles [14], beautiful for its simplicity as well as for its surprising conclusion. He treats Hypothesis B_+^n for $n > 2$ under the additional assumption that all of the fields of definition of the groups $P_i/O_p(P_i)$ have order at least 5. (Some smaller fields are handled under additional hypotheses). He proves that G has a rank n (B,N)-pair. Under the additional assumption that G is finite, Tits [23] has shown that G is, in fact, a rank n Chevalley group. Hence, we are left with the small fields where, not surprisingly, there are examples of sporadic configruations.

The first paper to handle a "small field" was by Chermak [3]. He considered special cases of Hypothesis B_+^3 represented by the diagrams

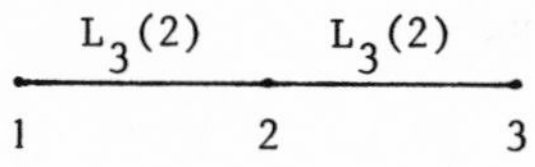

and

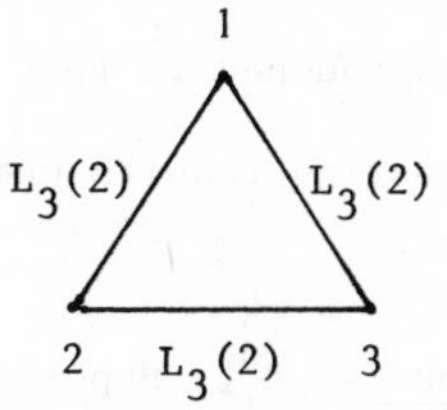

He shows that the group G of the first diagram is locally isomorphic to $L_4(2)$ and that the second diagram leads to a contradiction.

This work was extended by Timmesfeld [21] who treated arbitrary diagrams whose edges are labeled by perfect central extensions of $L_3(2^{n_{ij}})$. Substantial further contributions have been made by Stroth [19] and Timmesfeld [22] in yet unpublished results.

A complete analysis of Hypothesis B_+^n seems to be very difficult and that of Hypothesis B^n not even accessable. But we hope that the techniques we have presented here will inspire interest in some of these questions.

References

1. B. Baumann, Über endliche Gruppen mit einer zu $L_2(2^n)$ isomorphen Faktorgruppe, Proc. Amer. Soc., 74(1979), 215-222.

2. N. Blackburn, B. Huppert, Finite groups II, Springer Verlag 1982.

3. A. Chermak, On certain groups with parabolic-type subgroups over $\mathbb{Z}_2$, J. London Math. Soc., (2) 23(1981), 265-279.

4. A.L. Delgado, On edge-transitive automorphism groups of graphs, PhD-Thesis, Berkeley (1981).

5. P. Fan, Amalgams of prime index, PhD-Thesis, Berkeley (1984).

6. P. Fong, G. Seitz, Groups with a (B,N)-pair of rank 2, I, II, Invent. Math., 21(1973), 1-57, 24(1974), 191-239.

7. D.M. Goldschmidt, Automorphisms of trivalent graphs, Ann. of Math., 111(1980), 377-404.

8. D. Gorenstein, Finite groups, Harper & Row (1968).

9. G. Higman, Odd characterizations of finite simple groups, Univ. of Michigan (1968).

10. B. Huppert, Endliche Gruppen I, Springer Verlag, New York (1967).

11. F. Levi, Über die Untergruppen der freien Gruppen, Math. Z. 37(1933), 90-97.

12. R.P. Martineau, On 2-modular representations of the Suzuki groups, Amer. J. Math. 94(1972), 55-72.

13. R. Niles, Pushing up in finite groups, J. Algebra 57(1979), 26-63.

14. R. Niles, Finite groups with parabolic type subgroups must have a BN-pair, J. Algebra 75(1982), 484-494.

15. K.H. Schmidt, Punktstabilisatoren auf viervalenten Graphen, Diplomarbeit, Bielefeld (1984).

16. B. Stellmacher, On graphs with edge-transitive automorphism groups, Ill. J. Math., 28(1984), 211-266.

17. G. Stroth, Graphs with subconstituents containing Sz(q) and $L_2(r)$, J. Algebra 80(1983), 186-215.

18. G. Stroth, Kantentransitive Graphen und Gruppen vom Rang 2, J. Algebra, to appear.

19. G. Stroth, Parabolic systems over GF(2) whose diagrams contain double bonds, I, Preprint 169, Freie Universität Berlin.

20. M. Suzuki, On a class of doubly transitive groups, Ann. Math. 75(1962), 105-145.

21. F.G. Timmesfeld, Tits geometries and parabolic systems in finitely generated groups, I, II, Math., Z., 184(1983), 377-396, 449-487.

22. F.G. Timmesfeld, Amalgams with rank 2 groups of Lie-type in characteristic 2; Preprint Gießen (1984).

23. J. Tits, Buildings of spherical types and finite BN-pairs, Lecture Notes in Math. 386(1974), Springer Verlag.

24. K. Zsigmondy, Zur Theorie der Potenzreste, Monatshefte für Math. & Phys., 3(1882), 265-284.